Plastic Films for Packaging

Technology, Applications and Process Economics

CALVIN J. BENNING

Plastic Films For Packaging

Technology, Applications and Process Economics

By Calvin J. Benning

851 New Holland Avenue
Box 3535
Lancaster, Pennsylvania 17604

Printed in U.S.A.
Library of Congress Card No. 82-51065
I.S.B.N. 87762-320-1

The Author

Calvin J. Benning received his Ph.D. from Ohio State University and has attended courses for Senior Managers at Dartmouth and the IRI Research Managers Course at Harvard. He was part of the team that synthesized the first polycarbonate liquid prepolymers and produced the first one shot polyurethane foam process in the U.S.

Eleven years at W.R. Grace were devoted to Polymer Applications where he supervised the development of: a new class of shrink films, plastic foams, porous membranes, rigid containers and specialties. As Director of Wood Products and Industrial Packaging at International Paper, he continued to expand his expertise in the new product area where he managed groups that developed 'rigid-when-wet' corrugated containers, resins and adhesives for wood products, and special polymers. He is the recipient of the first International Award in PLASTIC FOAM for his 2-volume treatise in 1973. He has approximately 46 U.S. patents and publications. Dr. Benning founded and was the first chairman of the Gordon Research Conf. dealing with Plastic Foams. He is on the Board of Trustees of the Plastic Institute of America and has given courses and lectured for the ACS, PIA and several colleges. Dr. Benning is currently, as of 1979, Chief Environmental Officer of Essex Chem Corp.

Introduction and Overview

The data contained in this treatise have been derived from many sources. Originally this work started as a shrink film study. Later it was expanded to include an indepth comparison of "shrink vs. stretch" (a techno-economic comparison) and a technological look at plastic laminated film structures in packaging. This final update includes a look at major packaging films and the impact of changing technologies upon monomer supply, resin cost and supply, and plastic film markets in packaging.

Several key points bear particular emphasis at this time. These are: (1) the emerging technology of linear low density polyethylene and how this new resin and the versatile, low cost processes for polymerization (relative to LDPE) will affect market trends, (2) the effect of decreasing demand for vinyl chloride and styrene monomers on the ethylene monomer glut (this includes the decreasing importance of PVC as a food packaging film), and (3) the persistent problem of operating capacity of petrochemical plants over supply and new production capacity.

All of the data in this manuscript comes from the various sources listed in the many different literature references in Chapter 13. To give exact credit to each and every author is almost impossible since many references present similar information. Although the conclusions and remarks are the authors, similar conclusions in some cases have been voiced by others and credit is shared with them.

There are two main objectives in this effort: first, to define the impact that new technology in polymer science and raw material supply patterns will have during this decade on market size, share of market, and the economics of flexible packaging films. Second, to define and explain the science and technology involved during the transitions between monomer, polymer, packaging film, and application.

The first section describes market trends and uncertainties. The second and third sections describe the theories and technology of two major areas: shrink and stretch films. This is followed by a description and analysis of individual polymeric systems as each pertains to packaging films, oriented film products, and laminated and coated structures. Polymer structure, as it directly influences application, is discussed in each resin section. Section six examines film laminates and, as noted, these laminates have some very interesting new thrusts based on specific properties, tailor-made to the application.

Contents

Introduction and Overview vii

Chapter 1
Economics, Markets and Supply Trends of Plastic Packaging Films 1
Market Trends 1
Monomer Supply 3
Polymer Supply (Markets) 6
Domestic Packaging Film Markets 7
Economics 11

Chapter 2
Oriented Packaging Films 15
Historical Development 15
Theory of Stress Induced Orientation 16

Chapter 3
Orientation Techniques 19
General Approaches 19
Crystallin Polymers 21
Bubble Process 23
Tenter Process 24
Process Comparison 28
Non-Crystallin Polymers 28

Chapter 4
Technology of Commercial Shrink Films; Stretch Films and Laminates 31
Vinylidene Chloride Polymers 31

Polystyrene (OPS) 33
Polypropylene (OPP) 37
History and Overview 37
Mono-axial Orientation 39
Biaxial Orientation 39
Resin Properties 46
Molecular Weight 46
Polymer Structure 47
Additives 47
Metallized OPP Film Laminates 48
Commercial Significance of OPP and its laminates 49
Background 49
Uncoated OPP Films 50
Modified Uncoated OPP 50
Saran Coated OPP Films 51
Heat Seal Coated OPP 51
Acrylic Coated OPP 51
Markets for OPP Laminates and Products 51
Future of OPP 54
Poly (ethylene terephthalate) PET 55
Market Overview 55
Technology and Properties 56
Application of Polyester Films and Laminates in Packaging 57
Vinyl Chloride Polymers and Copolymers 60
Overview 60
Processes and Applications 61
Methods of Sheet and Film Manufacturers, a comparison 65
Irradiated Polyethylene 67
High Density Polyethylene (HDPE) 68
Markets and Supply 68
Shrink Film Potential 69
New Developments 71
Background and Motivation 72
New Materials 72
New Equipment 73
Summary and Future 74
Low Density Polyethylene and Copolymers 74
Low Density Polyethylene 74
Overview 74
Structure and Processability 76

Package Integrity 78
Linear Low Density Polyethylene 80
Historical Development 80
Structure and Properties 82
New LLDPE Technology 85
Impact of Technology 85
Applications 88
Conversion Costs and Efficiency 89
LDPE/Vinyl Copolymers 92
EVA Stretch Films 92
Ethylene–Methacrylic Acid Copolymers 93
Ethylene–Methacrylate 95
Rubber Hydrochloride 95
Polybutene (OPB) 96
Nylon Film (ON) 98
Metallized Nylon Film Laminates 97

Chapter 5
Properties of Heat-Shrinkable Films 99
Degree of Shrink 101
Shrink Tension 102
Shrink Temperature 102
Shrink Mechanisms 104
Amorphous Polymers 104
Crystallizable Polymers 107
Orientation effects 108

Chapter 6
Film and Laminates and Their Markets 111
General Markets 111
Polypropylene (OPP) 112
PVC Film 113
Polyethylene Films 114
Polystyrene (OPS) 115
Other Films 115
Engineered Films 116
Relationship Between Polymer and Film 117
Produce 118
Fresh Meat 119
Fresh Fish 123
Cured and Cooked Meat 124

Frozen Foods 125
Dairy Foods 126
Bakery Goods 126
Bags and Sacks 127
Boxed Goods 132
Bundling 133

Chapter 7
Shrink vs. Stretch (A Problem of Economics and Energy) 135
Overview 135
Today and Tomorrow 136
Shrink Packaging 137
Wrappers 137
Shrink Tunnels 137
Stretch Wrapping 137
Advantages and Disadvantages 138
Stretch Film Packaging 139

Chapter 8
Non-Packaging Applications 141
Construction 141
Industrial Chemicals 141

Chapter 9
Coextrusion 143
Technology 143
New Materials 144
Markets—Applications 145
Economics 147

Chapter 10
Comparison of Commercially Available Shrink Films 149
Shortcomings of Shrink Films 151
Cost 151
Machinability 151

Chapter 11
Economic Evaluation of Shrink Film Processes and Products 155
Amortized Costs 155
Material Costs 156
Labor and Overhead 156
Services 156

Chapter 12
Future Outlook 159
General Comments 159
Raw Materials 161
Supply and Costs 162
Capacity and Growth 163
Technology Changes 165
Packaging Systems 166
Speed 166
Energy 167
Cost 167

Appendices
Bibliographies 169
Recent Packaging Film Articles 170
Reference Books, Monographs and Proceedings of Major Conferences 173
Historical Patents 174
European Packaging Applications by Market and Category 175
Shrink and Stretch Film Properties 176
Shrink and Stretch Film Suppliers (US) 177
Suppliers of Orientation Equipment 180
Suppliers of Stretch Wrapping Machinery 181

Chapter 1

Economics, Markets and Supply Trends of Plastic Packaging Films

Market Trends – The market thrusts of the various polymeric films can be described from the viewpoint of the resin properties, cost, and availability, in other words, its competitive position in the market place. Predicting market thrust (or penetration) is complex, especially when one considers the impact of outside forces such as:

a) The emerging expertise that allows one to make LDPE, HDPE, PP, and LLDPE (with approximately 10% of the cost of a new facility) at the same facility (i.e. dual purpose reactors etc.).
b) The high cost of energy and labor in assembling pallet loads and bundles for shipping.
c) Environmental constraints on vinyl chloride and PVC manufacturing facilities; (i.e. air and hazardous waste controls) and plasticizers like DEHP.
d) The emergence of new petrochemical and polymer facilities in Alaska, Canada, Mexico, and the Near East (all near natural gas or oil sources).

Four film areas should continue to maintain steady growth. LDPE which includes LDPE and LLDPE (approximately 4% year); high density polyethylene (HDPE) (9-10% year); polypropylene (6-7% year); and polyester (7-8% year). PVC packaging film and cellophane will continue to lose position in food packaging. (Figure 1 and Figure 2). Most of the growth in PVC films is expected to be in non-food packaging and non-packaging areas.

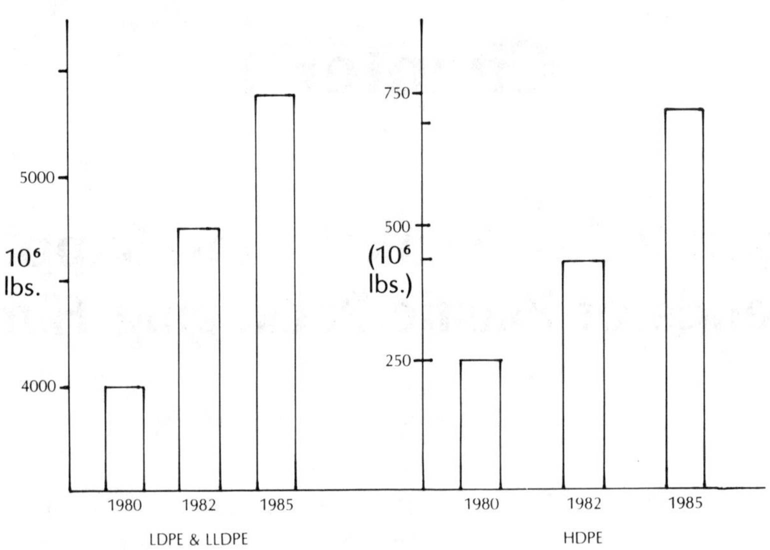

Figure 1. LDPE, LLDPE & HDPE market demand (million lbs) for film and sheet. (Source: Dow Chem & Plastics World (Nov. 1981, p. 13)).

The advent of LLDPE will make up the bulk of the growth in polyethylene films used in packaging. This intrusion will take place where clarity is not an essential property and where abuse resistance is a key – namely in material handling and bags. Some experts believe LLDPE will capture the grocery sack market and replace Kraft paper. If any system accomplishes the goal, it will probably also involve HDPE or blends of HDPE because of HDPE's rigidity and lower elongation.

Polypropylene could be in short supply in 1984 just prior to new capacity "coming on stream." PP's sustained growth in the overwrap market and in the baked goods market is predicted due to the lower "applied" cost of OPP and the growth of OPP as a laminate

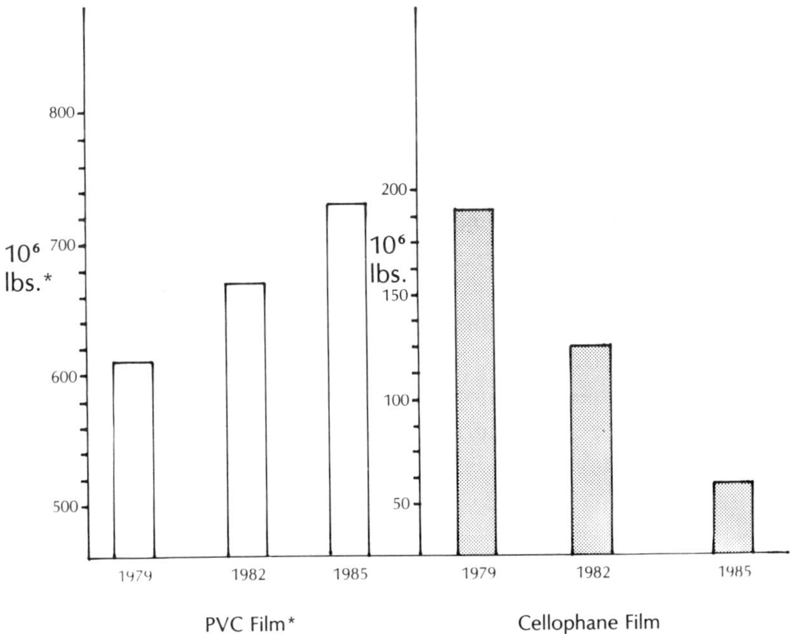

Figure 2. Film Market Projections. (Ref. Plastics World Dec. 1981).
**When comparing PVC film markets and growth (based on lbs.) one should use data that allows for differences in yield – i.e. m²/kg. Most of the growth in PVC film is in non food packaging areas.*

base, specifically as a lower cost competitor to oriented polyester, nylon and cellophane. Cellophane is expected to lose most of its market position as an overwrap.

High density polyethylene can be expected to continue to replace glassine in the "bag-in-box" application (dry mixes, crackers, cereal, etc.) and merchant type bags (and grocery sacks as noted before).

Polyester film (together with OPP, HDPE, and oriented nylon) will continue its market growth in the laminated, structured and specialty films area.

MONOMER (SUPPLY)

Basic petrochemicals will be in oversupply for some time. Between 1980 and 1982 relatively little growth is seen due to a significant increase in plant capacity and raw material costs. This means low profit margins. These problems will continue because of the planned decontrol of natural gas, which will provide additional costs to U.S. producers and increased foreign competition. Therefore, poor profitability, excess capacity, and increasing feed stock costs will persist.

Perhaps the best word to describe the situation is "chaos" because, in addition to these problems, one can expect a flood of petrochemical derivatives between 1985 and 1987 from outside the U.S. New imports will come from large, ultra modern, automated, and efficient facilities constructed close to the gas fields or well heads. The only positive note is that these new plants are expensive and are using expensive capital; however, raw material costs in Mexico, Canada and the Middle East are quite low and could offset the high cost of capital. These disruptions come at a time of low growth and undercut market forecasts.

A Chemical and Engineering News article* notes that the total production of ethylene, propylene, benzene, butadiene and p-xylene in 1982 will be 62 billion pounds, the same as 1979.

The feedstocks to watch are ethylene and propylene. In 1979 the industry was almost at capacity but in 1981 the sluggish economy, together with conservation, had reduced production of ethylene to 70% of capacity. Propylene is in a better position. Tighter supplies have helped to maintain prices and in the first half of 1981 resin prices increased 18 to 23% (Figures 3 and 4).

An example is Gulf Oil's activation of its 800×10^6 lb. capacity monomer plant in Cedar Bayou, Texas.

It is difficult to predict where propylene monomer prices will go, except up. Other contributing factors are: a shift to lighter natural gas liquids in the ethylene crackers, instead of heavy naphthas. This means less propylene by-product. In addition, propylene is being used with isobutane, (currently selling for approximately 90-96¢ gal.) to make a low-cost octane booster for gasoline. The decrease in demand for styrene and vinyl chloride monomer also contributes to inventory problems, i.e. a surge in ethylene inventory.

No new source for feedstocks will arise unless coal tar chemicals receive more attention.

Cellulose feedstock will continue to be directed towards making paper or wood products. In addition, emphasis will increase within the forest products industry to make it energy self-sufficient and put the industry in a cogeneration mode. However, the cost of paper products will

---

*CHEMICAL & ENGINEERING NEWS – November 1981, p. 11, Bruce Greek and W. Fallwell.

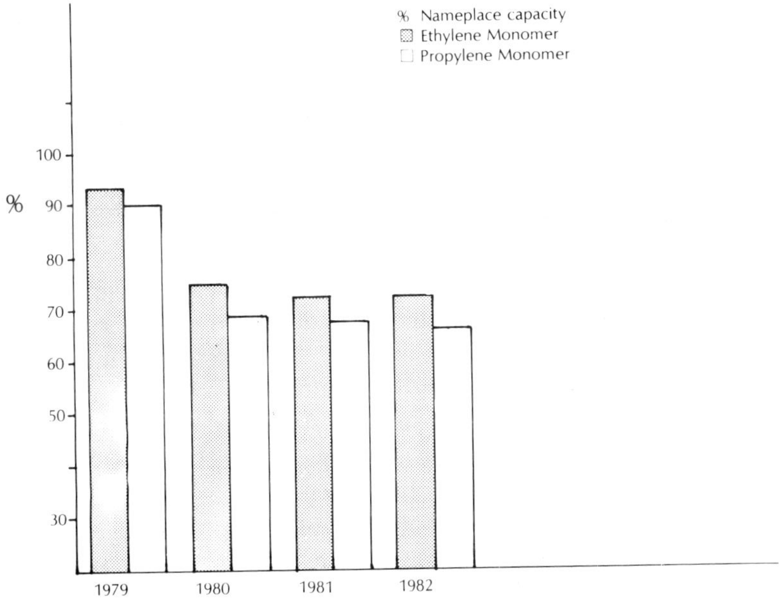

Figure 3. Production as a percent of nameplate capacity for ethylene and propylene monomer. Source C&EN estimates Nov. 30, 1981 and Plastics World.

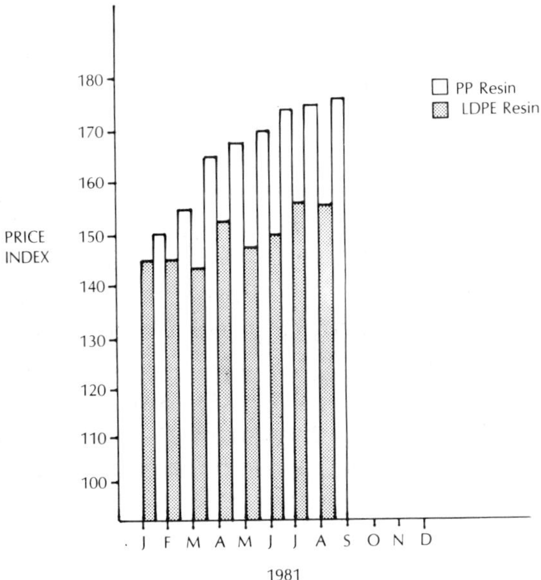

Figure 4. Price Index for PP and LDPE. Dec-75 = 100. Source Plastics World Data

continue to grow at the same general rate as the consumer price index, plus the added cost of environmental control. It's interesting to note that polyolefins are used in the manufacture of gallon and half-gallon milk containers. Although it's evident that paper is no longer the "cheap alternative," its strength, modulus and low extensibility help guarantee its position in packaging.

POLYMER SUPPLY. (MARKETS)

The polymer industry will be out of a polypropylene oversupply situation by 1982, when it is predicted that demand will approach 90% of nameplate capacity. Figure 5 shows that 1983-1984 will be critical years for polypropylene supply. We can expect that prices will maintain a high level and that margins will increase (note the trend indicated in Figure 6). We may expect a steeper price increase as producers try to recover some of the losses encountered by high monomer cost and poor price stability in 1977-1979. Industry officials expect polypropylene sales in 1981 to be approximately 15,000 tons more than in 1979 (in spite of business fall-off) (Figure 7).

Polyethylene is a different story. LDPE sales are approximately 57,000 tons below the 1979 high. Prices are decreasing due to an over capacity of 2.3 billion pounds and the intrusion of LLDPE (Figure 7 and 8). One could reason that the advent of LLDPE technology and the ability to convert existing facilities from LDPE to LLDPE at 10% of the cost of a new facility could change this. However, it may be wishful thinking. Because of down gauging, LLDPE will replace LDPE at *lower cost* and *weight*. If this logic is proper, one could expect that polyethylene overcapacity will continue and margins for conventional LDPE will continue to deteriorate. What we will see is a dramatic switch in the film market from LDPE to LLDPE.

For example, if LDPE (both high pressure and linear low density) is to maintain a 4% growth rate and LLDPE is expected to maintain a significant growth (note that the estimated 25% growth rate for 1980 to 1983 is due to the small base), then we can expect chaotic pricing of LDPE film grade resins and some significant facility closings or switching to LLDPE production. (Union Carbide stopped making LDPE at Texas City and Torrence facilities and the Penuelae Plant stopped production, others are expected to follow.)

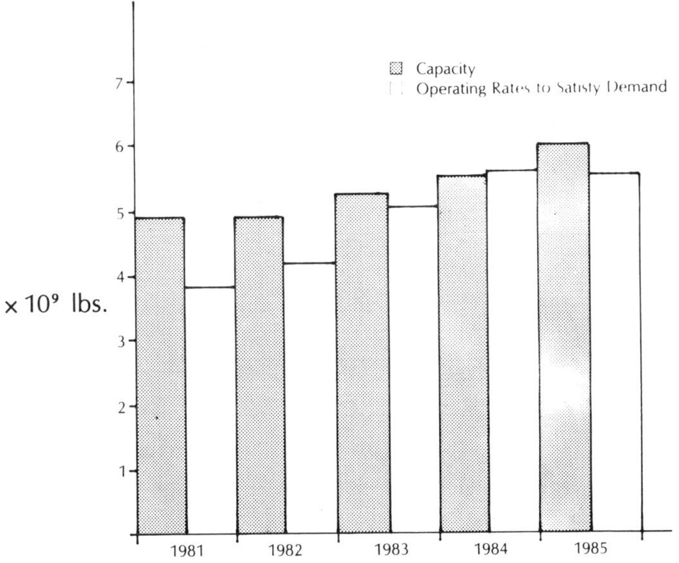

Figure 5. "Polypropylene supplies will be adequate but watch 1984.
Sources: Northern Petrochem; Plastics World.
Demand:
1981–82-84% of nameplate capacity
1982–90-92% of nameplate capacity
1983–98-100% of nameplate capacity
1984–100-105% (very tight market)
1985–94-95% of nameplate capacity.

DOMESTIC PACKAGING FILM MARKETS

According to reports put out by Predicasts and summarized in several issues of Plastic World (Nov. '81-Dec. '81) the packaging film shipments are falling very short of consistently optimistic industry predictions. A summary of plastic film shipments (in million lbs./yr.) plus the predicted shipments in 1995 give an interesting view of the industry (Figure 9).

However, a close analysis of the data is warranted:

> Cellophane is under attack by OPP films as an overwrap and as a laminate base. Also, the cellophane process is expensive due to environmental constraints in the areas of water quality, air and hazardous waste disposal. Therefore, one would expect Cellulosics to continue to decrease. (Shipments are expected to drop at the rate of over 15%/yr. for the next 10 yrs.).

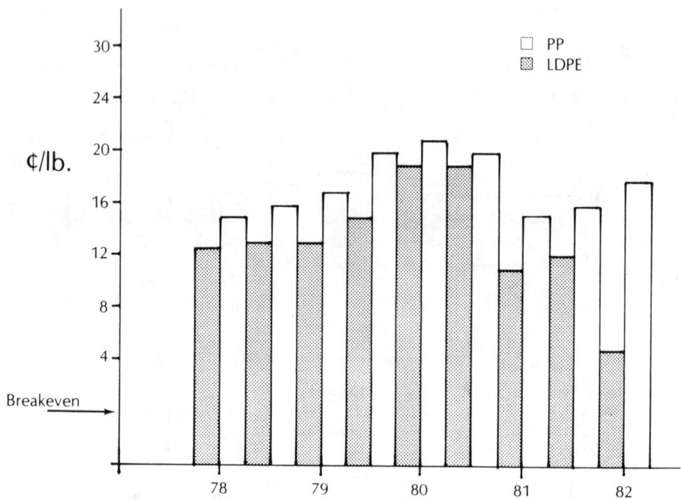

Figure 6. Difference between monomer cost (avg. ¢/lb) and polymer selling price (avg. ¢/lb).
Source: Plastics World

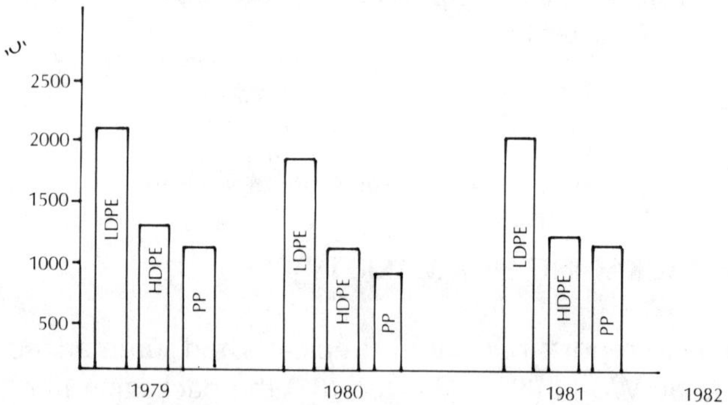

Figure 7. Recovery of sales is slow from 1979 high. (10^3 metric tons).*
PP–1981 sales = 1979 + 15,000 metric tons
HDPE–1981 sales = 1979 – 30,000 metric tons
LDPE–1981 sales = 1979 – 57,000 metric tons
Source: Plastics World

PVC films are on the increase. The prediction is an increase of approximately 350 million pounds between 1979 and 1995. This pertains to non-packaging or non-food packaging products such as windows and wallpaper, and in the areas of

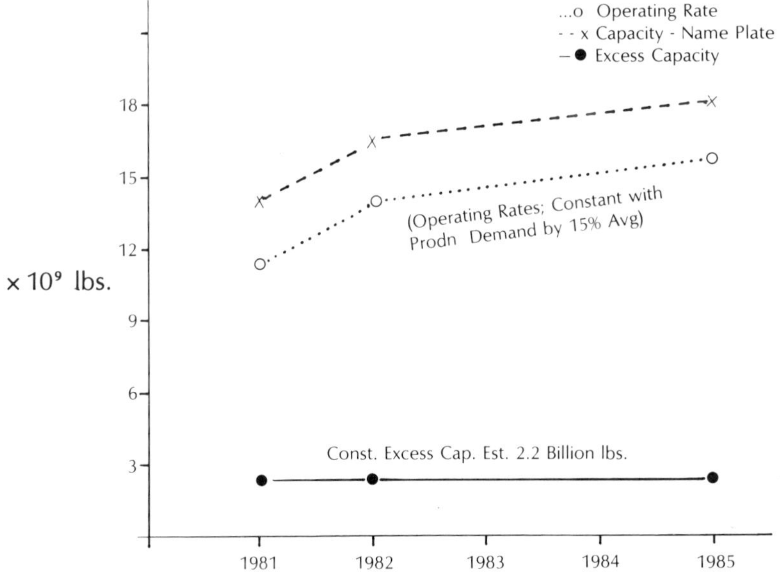

Figure 8. LDPE, LLDPE, HDPE production. Source: Dow Chemical; Plastic World.
PE–Operating Rates 1980–83-85%; 1981–84-86%; 1985–86-87% (est.)

LDPE –	*7793*	*7291*	*7600*	*8000*
HDPE –	*5010*	*4405*	*4600*	*4950*
	79	*80*	*81*	*82*

10^6 lb.

housewares and construction. The competitive stature of PVC in the food area continues to be eroded by PE and PP film. The total market for PVC films will grow at 5 to 6% year.

HDPE film will grow at almost three times the rate of PVC and LDPE – or almost 10% between 1979 and 1995, mainly due to its rapid growth in the merchant bag area and replacing glassine in "bag-in-box" applications.

LDPE (including LLDPE) will maintain less than moderate growth in the food area, and good growth in stretch film applications and shrink film packaging of industrial domestic products.

Industry reports expect the total domestic polyethylene packaging

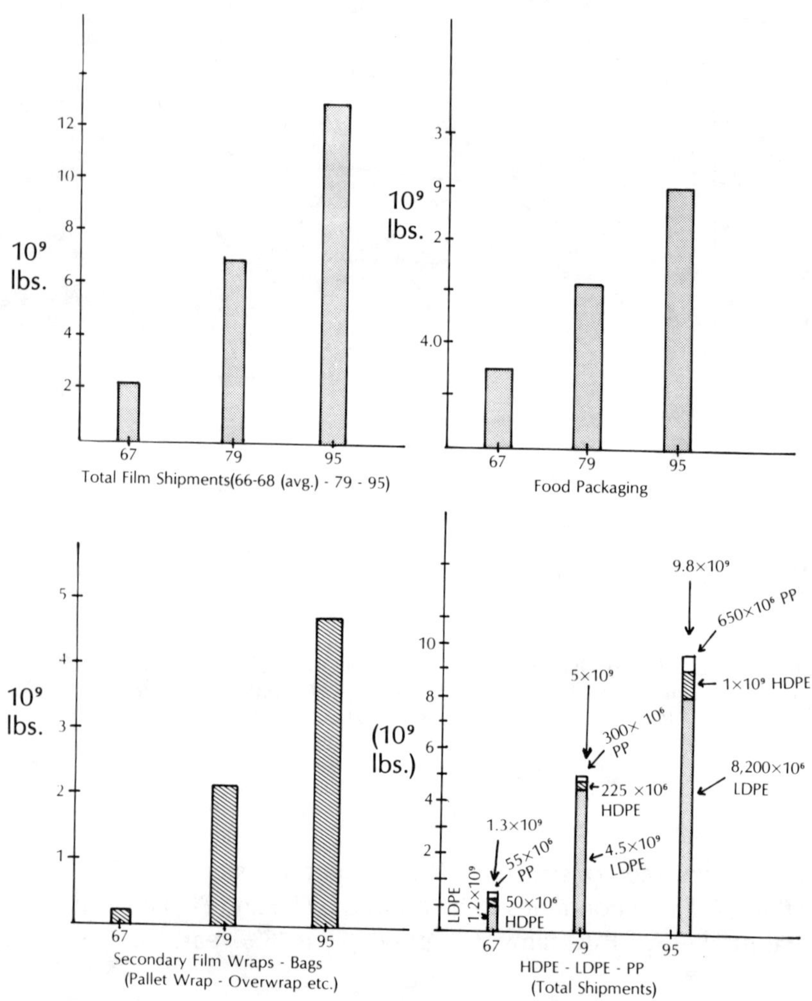

Figure 9. Source: Plastics World, Dec. '81 - Predicast Report. Note: 1967 and 1979 are actual figures; 1995 is an estimate.

film market – this includes food, material handling and indirect packaging (i.e. bags and overwraps) – to continue a sustained growth. According to optimistic producers this would amount to 7-8% year; according to a Predicasts study (or noted in Plastics World, Dec. '81 issue) an average of 4-5% year is more realistic. The real growth is in off-shore or non-U.S.A. production and this will affect the U.S. exports and some local upsets.

Two PE film areas look extremely interesting – the linear low density polyethylene process and the resins derived from it (the ethylene, alpha-olefin copolymers), and high density polyethylene. LLDPE and blends will affect areas where clarity is not a requirement (i.e. bags and stretch wrap). High density PE will continue to penetrate the bag-in-box market and replace glassine and wax paper.

The OPP area also presents a challenge. Latest calculations show that oriented polypropylene film can be produced at 50% above raw material costs and that there is still 30,000 tons per year of the cellophane market available.

In addition, new applications, such as packaging tapes, are being test marketed. Corporate confidence in OPP is also increasing as evidenced by a 9,000 ton per year specialty OPP film facility being brought on stream by Northern Petrochemical in Streamwood, Illinois. The plant output will concentrate on coating grades for snack food film (to be used with PVC primers and adhesives).

Generally, the polyolefin packaging films will continue to be the workhorses of the industry.

ECONOMICS

The previous sections have shown the impact of LLDPE on price and the decreasing margins associated with LDPE due to over-capacity and increasing competition from lower cost LLDPE. All this is at a time of decreasing overall demand.

The loss of some LDPE markets can be directly attributed to downgauging and the ability to use lower cost resins. Figures 10 and 11 give an excellent view.

Competition is also sharper in laminated films. Three competitive laminates are compared below:

Cellophane	– \$0.10 per 1,000 in.^{2}
OPP	– \$0.064 per 1,000 in.^{2}
Coextruded (3 ply) (HDPE/MDPE and Sealing Layers)	– \$0.04 per 1,000 in.^{2}

So here is a case where coextrusion is producing equivalent film at lower cost by incorporating lower cost resins.

Another example is the new Hefty® trash bags – a 3 ply coextruded structure:

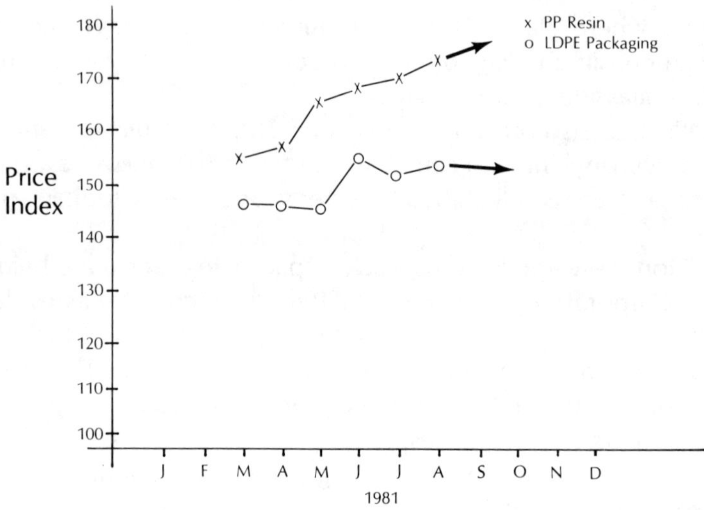

Figure 10. 1981 price indexes for packaging films. (Source: Plastics World, Oct. '81.)

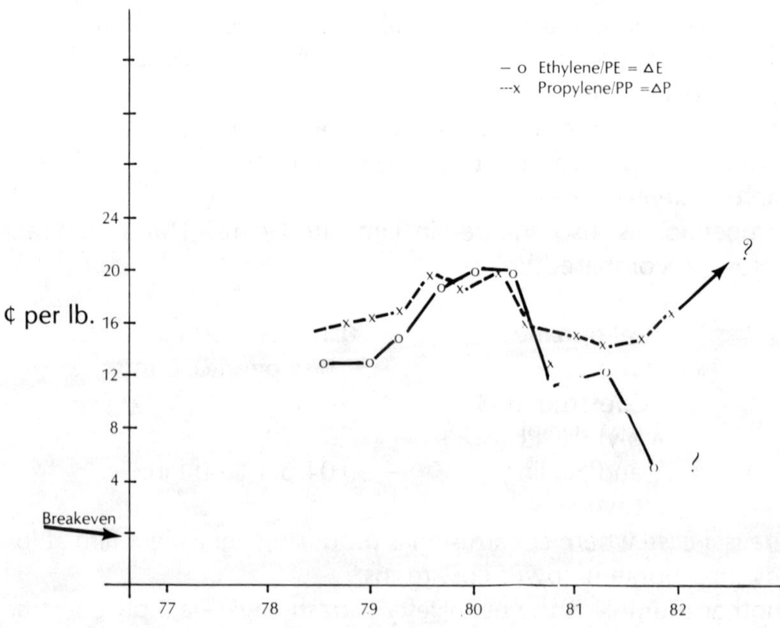

Figure 11. Difference between monomer cost (¢) and average polymer selling price between 1978 and 1981. (Source: Plastics World.)

1.5 Mil. -(3 Ply)- LLDPE/LDPE/LLDPE
replacing 2.0 Mil. of LDPE and
1.3 Mil -(3 Ply)- LLDPE/LDPE/LLDPE
replacing 1.5 mil. LDPE kitchen trash bags (both structures are replaced with significant savings in raw material cost).

Chapter 2

Oriented Packaging Films

HISTORICAL DEVELOPMENT

Biaxial orientation is a process whereby a plastic film or sheet is stretched in such a way as to orient the polymeric chains of the plastic parallel to the plane of the film. Biaxially oriented films possess exceptional clarity, superior tensile properties, improved flexibility and toughness, improved barrier properties, and the unique property of engineering shrinkability. Biaxially oriented Saran® (PVC-PVDC copolymer), PVC, rubber hydrochloride, radiation crosslinked PE mixtures of LDPE and HDPE, polystyrene, and polypropylene are the mainstays of the industry. In the food industry these are widely used where their specific shrink properties and barrier properties are particularly valuable, or, in the case of OPP, as a base film for laminar structures.

Virtually any thermoplastic material can be oriented. The first biaxial orientation process (called Luvitherm® by I. G. Farben, for plasticized PVC) was developed in Germany in 1935. During World War II oriented polystyrene film was made in Europe for capacitors and coaxial cable insulation. In 1936 natural latex was used in France to shrink wrap perishable foods. The film was stretched and allowed to shrink back on the goods to be packaged. This technique is still used with films like LLDPE blends, EVA, plasticized PVC, and LDPE/VAC copolymers. The process has been referred to as stretch wrapping, to distinguish it from heat shrinkable film packaging.

In 1948 Dow Chemical developed the Saran® resins and Dewey & Almy (now the Cryovac Division of W. R. Grace) developed the film process and introduced Saran® shrink film to the marketplace. This development opened the eyes of the marketplace profession to a new era in packaging.

Shrink wrapping in this sense began with the wrapping of poultry for deep freeze storage and has been dominated by Cryovac technology for many years. A film bag was slipped over the bird, a light vacuum was drawn and the mouth of the bag clipped shut. The bag and contents were then immersed in a hot water bath where the bag shrank tightly around the contents. Apart from giving an extremely neat package, shrink wrapping had the advantage of eliminating freezer burn (dehydration). (This technique is now used in many other food packaging applications.)

The major breakthrough in oriented high performance films came after Cryovac developed the Saran® bags for packaging frozen fowl. In the mid-fifties E. I. Dupont started development of high and low density polyethylene blends, and Cryovac was in the midst of commercialization of their radiation crosslinked polyethylene film. By middle 1960 Cryovac had developed a complete line of shrink films consisting of crosslinked PE, OPP, oriented polystyrene (OPS), and polyester laminates for the fresh and prepared food industries. Minnesota Mining & Manufacturing Company and Reynolds Aluminum Company have also been very active in the area of structured packaging films based on coated laminates for boil-in-bag applications. Reynolon® and Pliofilm® were very active stretch films. Since mid-1960, three polymers have emerged as the giants in the shrink or stretch film markets. These are PE (and its copolymers), PVC and PP. The phenomenal growth of OPP, to replace cellophane and to serve as a laminating base for barrier and special applications, indicates the main criteria for use – "*cost* and functionality."

THEORY OF STRESS-INDUCED ORIENTATION

Molecular orientation during stretching takes place in the following manner: Below their glass-transition temperatures (T_g), polymer chains are rigid. At the glass-transition temperature they become more flexible and are able to unfold as stress is applied. If a mass of randomly coiled and entangled chains is above T_g when stress is applied, as in biaxially stretching, the polymer chains disentangle, unfold and straighten, and also slip past their nearest neighbor.

There are three rheological components to this process: First, E_1, the instantaneous elastic deformation caused by bond deformation or bond stretching, which is completely recoverable when the stress is released (Hookian behavior). Second, E_2, the molecular alignment deformation caused by uncoiling which results in a more linear molecular arrangement parallel to the surface and which is frozen into the structure when the material is cooled (Elastic behavior). And third, E_3, the unrecoverable viscous flow caused by molecules sliding past one another. The elastic component, E_2, is the major component of the stretching process.

When the film is rapidly stretched at a temperature slightly above T_g, E_1 deforms instantaneously followed by E_2 deformation (Table 1). If at some time t_1, E_2 is large, relative to E_3 and the material is rapidly quenched, the alignment deformation E_2 is frozen into the structure. If the identical stretch and quench sequence is carried out at a significantly higher temperature, E_2, the desired alignment component of the deformation, will be a smaller proportion of the total deformation; this is mainly because viscous flow, E_3, increases as the temperature is increased, so that E_3, deforms more in time t_1, allowing E_2 to relax more during the time consumed by the stretching operation. Stretching is thus a dynamic process in which orientation and return to random coil (relaxation) occur simultaneously.

Table 1. Rheological Components of Orientation

E_1	Instantaneous elastic deformation caused by *Bond Stretching* – recovers when stress is removed.
E_2	Molecular alignment caused by uncoiling which is frozen into structure when cooled below T_g.
E_3	Nonrecoverable viscous flow caused by molecules slipping over each other.

On the basis of the above theory, some general rules for orienting polymers by stretching can be set forth.

1. The lowest temperature above T_g will give the greatest orientation (and greatest shrink strength, etc.) at a given percent and rate of stretch. E_3, viscous flow, is held to a minimum by keeping the temperature as low as possible.
2. The highest rate of stretching will give the greatest orientation at a given temperature and percent stretch. Since E_3 is a slower process than E_2, E_2 will predominate during rapid stretching.
3. The highest percent stretch will give the greatest orientation at a given temperature and rate of stretching.
4. The greatest quench rate will preserve the most orientation under any stretching condition.

These principles are illustrated in Figure 12.

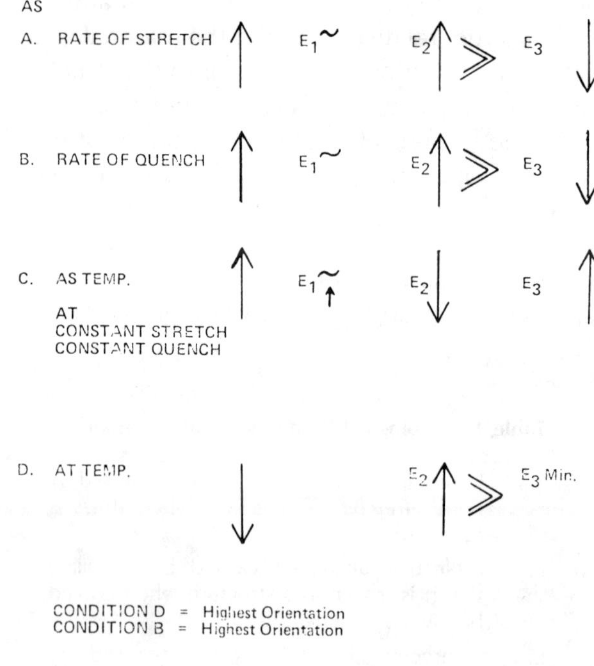

Figure 12.

Chapter 3

Orientation Techniques

GENERAL APPROACHES

The polymer sheet or tubing to be stretched or oriented can be made by the usual techniques, such as extrusion, calendering, or solvent casting. Extrusion is the most common method and either annular dies or slot dies may be employed depending on the raw material and the method of orientation. The economics of supply, cost of capital and manufacturing, etc., usually determine if calendering is the preferred method (i.e. of producing PVC film). The third approach, casting from a polymer solution, is becoming very unattractive due to the high cost of solvent, the high cost of capital, and a long list of environmental concerns. (Reynolds Company has a tetrahydrofuran solvent process for cast PVC.) Both calendering and solvent casting produce film with excellent thickness control compared to extruded sheet. However, as noted previously the economic and environmental problems have increased the operating costs of this process to such an extent that it is becoming unrealistic to operate.

Classical procedures of orienting polymers has been discussed in several British journals and books.*

---

*See Appendix 13.1.2

The orientation procedure for crystalline polymers has the following sequence:

(1.) Heat the polymer to above the melting point to destroy crystallinity (e.g. melt extrude).

(2.) Quench to minimize crystallinity and preserve the amorphous condition. (This will facilitate subsequent orientation.)

(3.) Reheat and orient by stretching at a temperature somewhat above the second order transition temperature but below the crystalline melting point.

(4.) Anneal (if desired) to reduce thermal shrinkage or rapidly quench to "freeze in" shrink energy. For this step the film is constrained from shrinkage during heat treatment.

For polymers of relatively low crystallinity, e.g. polyester, the procedure could consist of the following sequence:

(1.) Heat the polymer to a temperature well above the second order transition temperature so that the polymer is well into the viscoelastic region. This can be done with an extruder.

(2.) Cool the polymer to the proper temperature wherein the resin is elastic and orient by stretching at the prescribed temperature and rate. (Note: Here also one notes a decrease in strength at higher orientation temperatures, as in the case of polystyrene in Figure 13.

(3.) Anneal (if desired) to reduce thermal shrinkage. Again, the annealing conditions may be carefully controlled to prevent relaxation, and loss of shrink energy.

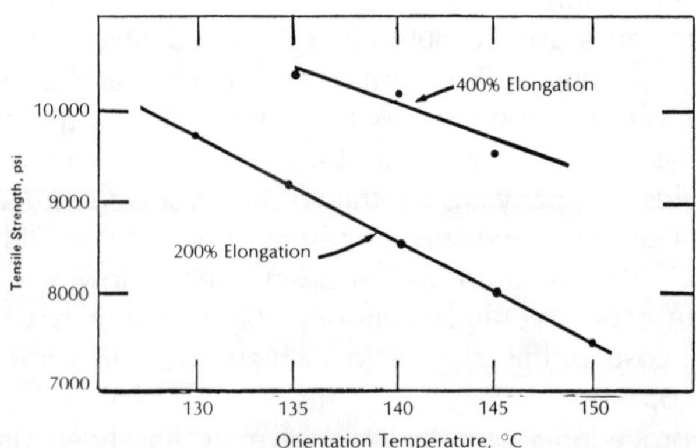

Figure 13. Enchancement of strength through orientation as a function of stretching temperature and percent elongation for polystyrene. (Courtesy SPE Journal.)

The specific orientation temperature will vary from polymer to polymer. It can be determined experimentally, or from the literature. Table 2 describes the second-order transition temperature, the crystalline melting point, and the orientation temperature range.

Table 2.*

Polymer	Second-Order Transition	Crystalline Melting Point	Orientation Temperature
Polyethylene terepthalate	70 C	255 C	85-110 C
Polyhexamethylene adipamide	45-50 C	250 C	67-75 C
Polyhexamethylene sebacamide	45-50 C	250 C	65-75 C
Polycaprolactam	45-50 C	250 C	65-75 C
Polyvinylchloride			
no plasticizer	105 C	170 C	115-145 C
5% plasticizer	90 C	170 C	100-130 C
15% plasticizer	60 C	170 C	70-100 C
Polystyrene	—	—	88-110 C
Polymethylmethacrylate	—	—	66-105 C
Polypropylene (d - .8825)		140 C	100-120 C
Polypropylene (d - .9015)		165 C	125-145 C
Polypropylene (d - .912)		180 C	140-160 C
Polyethylene (d - .92)		110 C	80-110 C
Polyethylene (d - .96)		134 C	120-130 C
Polyvinylflouride (d - 1.4)		193-198 C	175-185 C
Polyoxymethylene (d -1.35)		180-185 C	130-180 C

U.S. PATENT 3,231,643 (to Du Pont, January 15, 1966).

CRYSTALLINE POLYMERS

Most of the polymers used in manufacturing shrinkable films are crystallizable polymers. To obtain an orientable tape by extrusion, crystalline polymers must have minimum crystallinity. Therefore, it is necessary to quench as rapidly as possible to below the temperature of crystallization. The material should be cooled below its glass-transition temperature to ensure a low level of crystallinity, or should be cooled to a temperature at which the crystallization rate is low enough that no appreciable amount of crystallinity will develop in the material before it is stretched.

The amorphous or nearly amorphous material is then reheated to as low a temperature as practical above T_g and stretched biaxially very rapidly to prevent crystal growth. (This material can then be quenched to below T_g or to a lower temperature above T_g to give an essentially amorphous, readily heat-shrinkable film). (Table 3)

Table 3. Crystalline Polymers

1.	Extrude
2.	Quench (below crystalline M.P.)
3.	Reheat and/or orient
4.	Anneal (below crystalline M.P.)

Quenching is carried out either by extruding the web onto a chill roll or by passing it through a cold-water bath (Figure 14). Polyethylene terephthalate has a second-order transition point of 70°C and a minimum temperature for crystallization of 90°C, hence the resulting tape is stable. Films, such as polyvinylidene chloride and polypropylene, which have glass-transition temperatures below room temperature, show an appreciable crystallization rate even at room temperature. Accordingly, these films have to be oriented immediately after the formation of the tape, and this is essentially true with vinylidene chloride/vinyl chloride copolymers (Saran™).

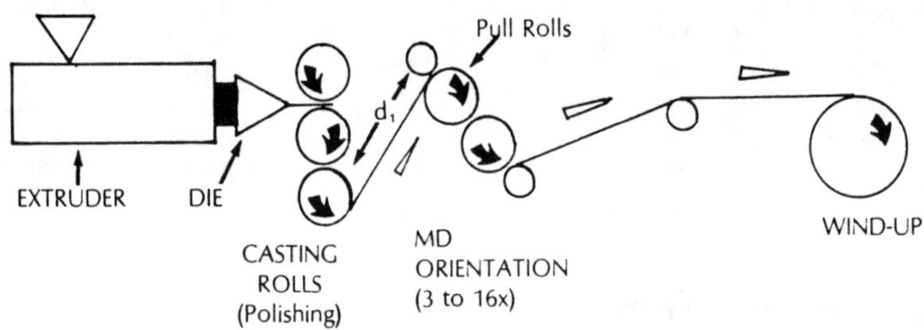

(*Note:* Distance d is small to diminish "necking" and improve web control.)

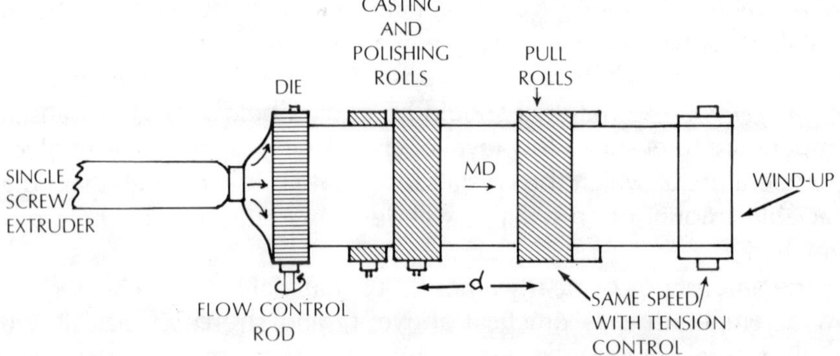

Figure 14. Schematic diagram of sheet extrusion and chill-roll casting unit. (Ref. - Plastics Films - J. H. Briston Halsted Press. Modern Packaging Films, Edited by S. H. Pinnir Publ. Butterworth, London 1967.)

BUBBLE PROCESS

The stretching of the polymer sheet or tape can be made either in the longitudinal or transverse direction; in a simultaneous or two-step operation. Two methods have been developed. One is called "bubble process" in which a supercooled tubular film is heated to orientation temperature and stretched by inflation between two sets of pinch rolls running at different speeds. The speed ratio of the rolls controls the orientation in the machine direction (MD) whereas transverse orientation depends on the ratio between final bubble and initial tube diameter. (Figure 15).

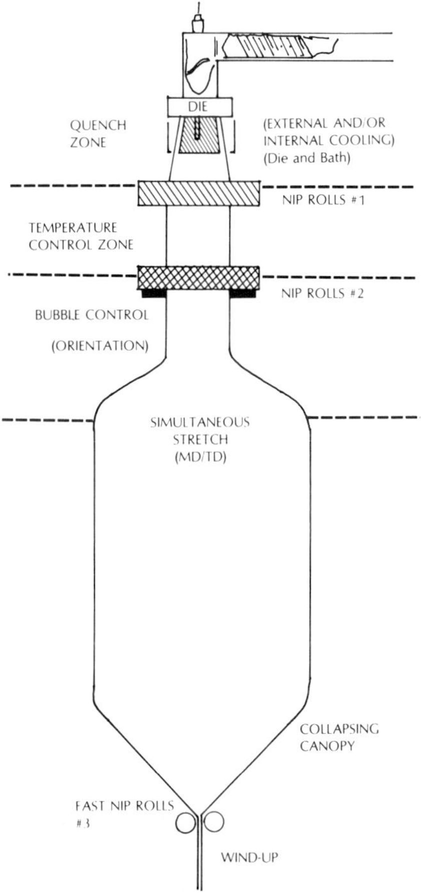

Figure 15. Orientation by the bubble process. Source: Modern Packaging Films, Edited by S. H. Pinnir Publ. Butterworth, London 1967.

For most oriented-film applications it is desirable to have a film as nearly balanced as practicable. This can be regulated by blow-up ratio and speed ratio of various draw rolls in the apparatus. The illustrations that follow show how one can regulate and balance properties by keeping the blow-up ratio (transverse stretching ratio) constant while increasing the speed of the take-off rolls (machine direction stretch). (Figure 16-19).

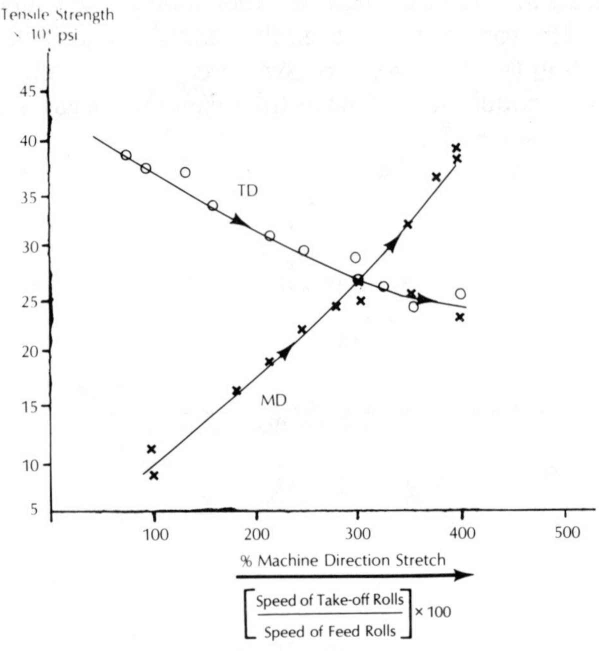

Figure 16. Oriented polypropylene by the bubble process.

TENTER PROCESS

In the second process, the tenterframe operation, the polymer is extruded through a slot die and quenched. Then the sheet is normally oriented in two steps. The first step is usually longitudinal orientation between rolls running at different speeds. In the second stage, the film enters a tenterframe, where it is stretched laterally by means of diverging chains of clips (Figure 20). Whereas the bubble process operates at constant pressure, the tenterframe process operates at a constant rate of elongation. Somewhat higher stretching forces are required in the second stage which may be carried out at slightly higher temperatures. This is mainly due to crystallization of the film during the first stretching operation. The tenterframe process can also be carried out as a simultaneous

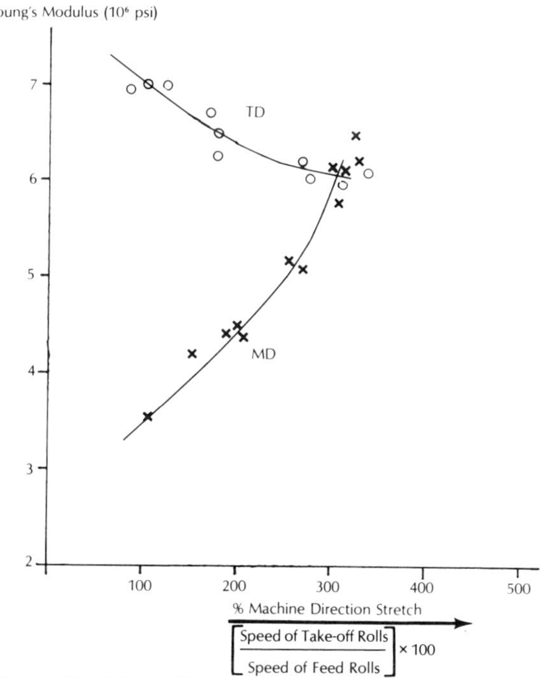

Figure 17. Oriented polypropylene by the bubble process.

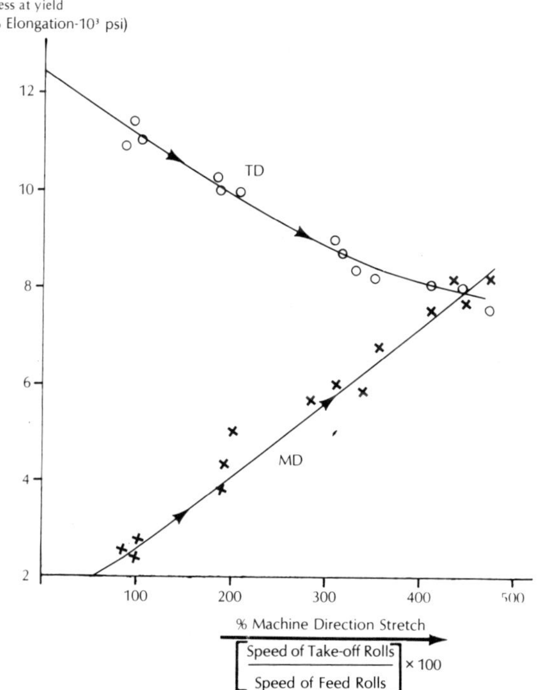

Figure 18. Oriented polypropylene by the bubble process.

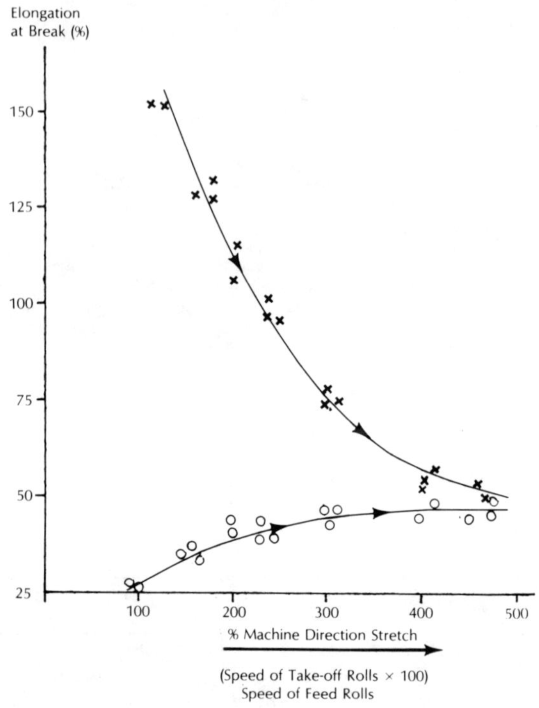

Figure 19. Oriented polypropylene by the bubble process.

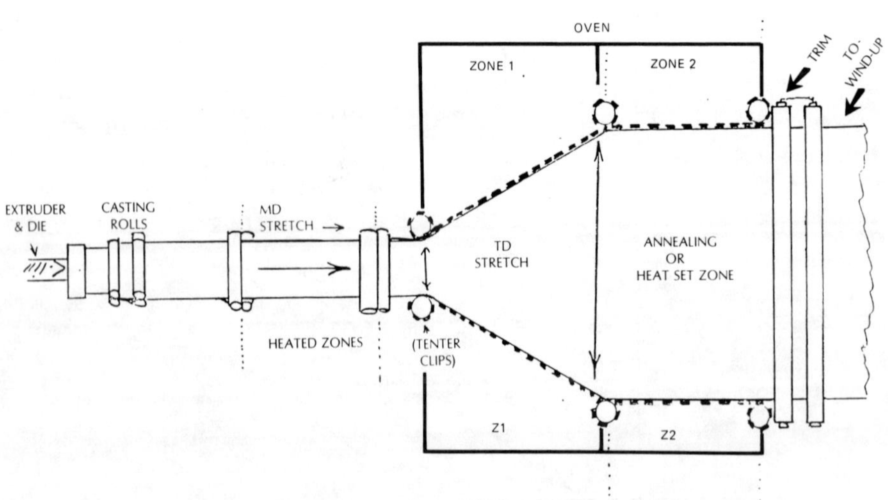

Figure 20. Two-step orientation by tenting (schematic diagram). Source: Modern Packaging Films, Edited by S. H. Pinnir, Publ. Butterworth, London 1967.

operation in which an extruded sheet with beaded edges is biaxially oriented in a tenterframe equipped with diverging roller grips for holding and stretching the film (Figure 21).

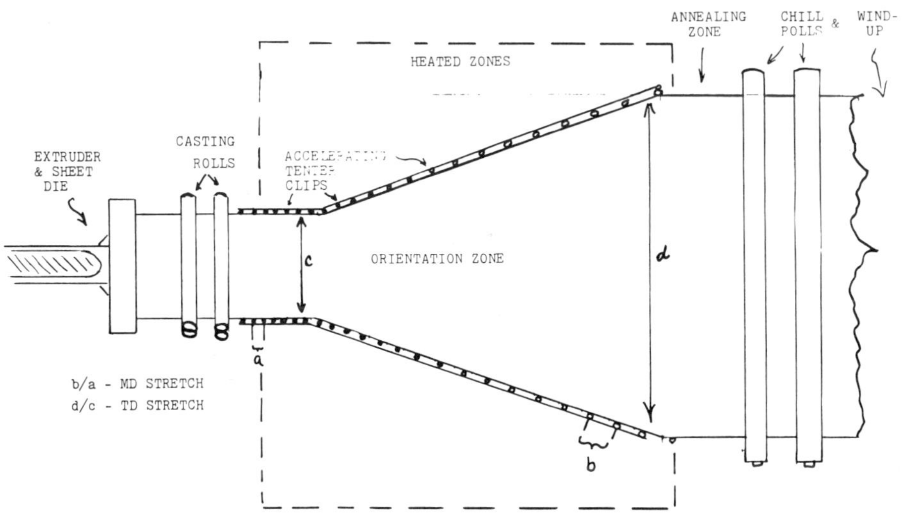

Figure 21. Simultaneous or one-step orientation (schematic diagram). Source: Modern Packaging Films, Edited by S. H. Pinnir, Publ. Butterworth, London 1967.

The tenterframe operation has the advantage of considerable versatility, producing films with a wide range of shrink properties. Both two-stage and simultaneous tenterframe procedures have been employed for orienting polyethylene terephthalate webs, and P.S. biaxially. (This process is required for very stiff materials having very low elongation at failure.)

After stretching, polymer alignment is locked into the film by cooling. When the oriented film is subsequently heated up to temperatures in the vicinity of the stretching temperature, the frozen-in shrinkage stresses become effective and the film shrinks. Strains and stresses which are related to the degree of orientation and the forces which were applied during stretching are thereby recovered.

To obtain a greater degree of heat stabilization, the stretched film is restrained from shrinking, heated to the temperature of its maximum crystallization rate (or higher), and held at this temperature until crystallization has occurred to the desired degree. It is then quenched to room temperature and the restraint on the film is released.

PROCESS COMPARISON

Advantages of the bubble process are:

1. Simultaneous biaxial orientation: process is controllable over a wide temperature range.
2. The bubble symmetry tends to produce even heating and is an important asset in assisting a uniform orientation in both machine drection and cross machine direction.
3. Edge scrap in the bubble process can be negligible since a simple slitting process can be employed.
4. Die design and symmetry is an established art for extrusion of heat sensitive materials.

Points in favor of tenting (two step orientation of flat film):

1. The melt stream is a continuous sheet; there is no seam as one sees in film dies (although these can be eliminated or minimized with vertical extruders and "straight through" dies).
2. Rheology of the melt and stretch behavior can be controlled in a two step orientation.
3. Quenching the film can be achieved more efficiently.
4. Film speeds are higher and gauge control is easily achieved in a tenting operation.
5. The tenting process produces a flatter film.

NON-CRYSTALLINE POLYMERS

Virtually all thermoplastics can be oriented, but one can orient amorphous materials more than crystalline materials. The stretching process frequently induces or increases crystallization particularly where there is some geometric regularity in the molecule. In the past, (prior to 1955) most of our attention was directed to the orientation of amorphous materials and polymers. With the commercialization of the Zeigler-Natta polymer technology, polyolefin research was directed more to the orientation of polyethylene and polypropylene and other low priced polyolefins.

The problem of crystallization is not significant with such polymers as polystyrene (and its copolymers), acrylics, polyesters, and polycarbonates since under normal conditions of polymerization and processing very little crystallization occurs.

Table 4. Illustrates the Process Steps Involved in the Orientation of Non-Crystalline Polymers

1. Extrude (or Calender)
2. Stretch (Orient)
3. Quench (and/or Anneal)

While PVC is not classified as a crystalline polymer, it is amenable to the manuacture of a relatively high shrink force film by the so-called double bubble process. This process has been used in Europe to produce OPVC. The process consists of an extrusion blown film exiting from the die, with higher than normal blow up ratio, followed by a second bubble of approximately two to one blow-up ratio, which puts the shrink force into the film. (Courtesy of Cryovac Div'n of W. R. Grace).

Author's note

At this point the reader is well aware that one can effectively orient polymeric films and sheet with relative ease. Although hot orientation puts in shrink, it results in very little, if any, shrink tension or force. Orientation at a lower temperature, near but below the crystalline melting point results in high shrink force. Chapter 4 describes the Technology of specific film systems used in packaging.

Chapter 4

Technology of Commercial Shrink Films, Stretch Films and Laminates

VINYLIDENE CHLORIDE COPOLYMERS

The principle of an early patent describing the orientation of Saran™ film by a bubble process is illustrated in Figure 22.

In this process an 85:15 vinylidene chloride-vinyl chloride copolymer, plasticized with 7% bis-(α-phenylethyl) ether, is extruded through a circular die orifice at 170°C. The tube is rapidly quenched in a water bath to render the polymer amorphous. After being flattened at the first set of nip rolls, the tube (approx. 25 mil thick) passes to a second set of nip rolls outside the cooling bath going at the same speed as the first set (10 ft per min.). Between the second and third nip roll sections, air is introduced into the tube to expand it to about 4X its diameter. This room-temperature stretching operation introduces orientation in the transverse direction. At the same time, the third set of nip rolls accelerates the tube by a factor of three or four, so that stretching in the machine direction is carried out simultaneously. The expanded tube (2 mil thick) is then collapsed and

TM – Trademark Dow Chemical Co., Inc.

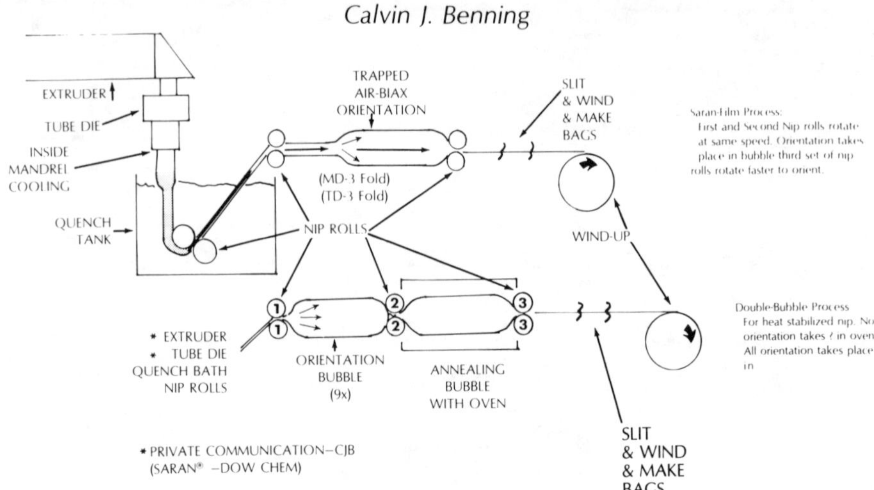

Figure 22. Preparation of Saran Film by Tube/Bubble Processes. The double-bubble process while proposed has not been commercialized. (Source: USP 2,452,080 to Dow Chem. 1948; Courtesy of Cryovac Div'n of W. R. Grace).

either wound as a tube for bags or slit and flattened to a single sheet. Table 5 compares the properties of shrink and bag grade films.

Table 5. Properties of PVDC/PVC (85/15) Copolymer Film[1]

Property	Shrink Grade	Bag Grade
Thickness (in.)	0.001	0.001
Yield (in²/lb.)	16,300	16,300
Tensile Strength (psi)	9,000	12,000
Elongation at Break	35%	25%
W.V.T.R.[1]	0.2	0.3
O_2 Transmission[2]	0.6	0.9

[1]gm/mil/24 hr/100 in² at 90% RH and 100°F
[2]c.c./mil/100 in²/24 hrs./atm. 0% RH (STP)
(Source: Ref. 13.1.9, Briston, Sweeting, Owsin, et al.)

As the tube of amorphous polymer is biaxially stretched, crystallization, (as well as orientation) takes place. Tensile strength increases as thickness diminishes. At a given internal pressure and temperature, the bubble diameter stabilizes at the point where the tensile forces equal the internal pressure. Further introduction of air into the bubble does not increase its diameter but only the length of the expanded portion. Bubble diameter can be altered by: (a) changing the orientation temperature; (b) changing the tube thickness; (c) preheating the tube before expansion; (d) altering the polymer composition or plasticizer content; and/or (e) changing the quench conditions to give more or less crystallinity in the cooled extruded tube. Thus several processing variables are critical to this particular process and must be precisely controlled if a uniform product is to be obtained.

The vinylidene chloride copolymer films, because of their impermeability to gases, are used mainly in food packaging. Cooked and spiced meats, frozen poultry, and cheese are wrapped to advantage in this film, as are frozen bakery goods. Indeed, shrinkable wrappings constitute a rapidly growing segment of the packaging industry, and one in which any heat-shrinkable oriented plastic film may find a place.

The heat stabilized form of vinylidene chloride copolymer film is often laminated to other films to obtain special properties. For example, it is laminated to polyethylene for better sealability; to poly (ethylene terephthalate) for greater toughness; or to other films or combinations of films where gas and moisture impermeability and dimensional stability are needed (Table 6). In addition to having the best available barrier properties at a reasonable price and thickness, Saran™ is also resistant to chemicals, resulting in its use as cap-liner material and for packages containing oily, greasy, or acidic substances.

Table 6. Shrinkage of Saran™ Films

	% Shrinkage 5 min in Boiling Water		–Thickness, mil –	
	MD	**TD**	**Before Shrinking**	**After Shrinking**
Unstabilized Saran™	30	40	5	12
Saran™ Stabilized Under Restraint at 100°C for 6 sec.	3	3	5	5

TM – Trademark Dow Chemical Co., Inc.
Sources: Cryovac "S" Film data and Dow Saran™ data.

POLYSTYRENE

Polystyrene was developed as a commercial material before the polyolefins. Liquid monomer polymerizes in bulk to a glassy solid which can be extruded and de-gassed in an extruder fed directly from the reactor and converted into free flowing beads. The polymer is atactic (noncrystalline). Extrusion of PS is fairly simple and straight forward. The typical properties of an extruded polystyrene film are

Thickness (in)	0.001
Yield (in^2/lb.)	26,200
Tensile Strength (psi)	9,000-12,000
Elongation at yield (%)	10-15%
Modulus	3 GN/m^2
Ref. 13.1.2 1 and 2	

Polystyrene film as with other polymers can be made more flexible by orientation. Polystyrene is one of the easiest polymers to orient (and to expand) since it melts at 100°C and the high melt viscosity allows it to be stretched up to 130°C. The barrier properties are poor after orientation.

The properties of 1 mil (25um) oriented polystyrene film (OPS) show it to be a stiff material. It can be heat sealed and cut by hot-wire; since shrink film wrinkles during heat sealing the film must be constrained (clamped) to give flat seals. (Polystyrene begins to shrink at 100C).

Polystyrene has been biaxially oriented by a variety of techniques – tentering, bubble, octagonal stretcher, and horseshoe mandrel processes. The tentering process is the most widely used process.

In the tentering process (Figures 23 and 24), a sheet of polystyrene is extruded from a slot die and passed around two or three polished chill rolls to reduce its temperature to 150°C. It next passes through the section where stretching takes place in the machine direction, introducing a three fold stretch in the machine direction. (If these two rolls are sufficiently close together, essentially no necking occurs.) The sheet may be cooled at this point or fed directly into the tenter frame. In this frame, a series of clips grasps both edges of the sheet. These clips are mounted side by side on endless chains. The tenter frame is divided into three sections. The first section, T_1, is for temperature conditioning of the material (105°C). The clips proceed parallel to one another. In the second section, T_2, (held at 105-130°C), the tracks diverge and cause the sheet to be stretched in the transverse direction. In the third section, T_3, the biaxially oriented sheet is cooled to below 85°C (T_g), the clips are released, and a 45 in. wide 10-mil film is obtained which has been stretched to three times its original dimension in each direction.

In order to provide certain minimum strength properties, the amount of stretch may be critical. For example, in the case of polystyrene, ASTM standards require a minimum orientation ratio of 2:1 in each direction. Lower ratios result in a brittle sheet. Orientation beyond 3:1 in each direction does not give appreciable improvement in properties; thus, 3:1 is considered optimum for general purpose polystyrene sheet.

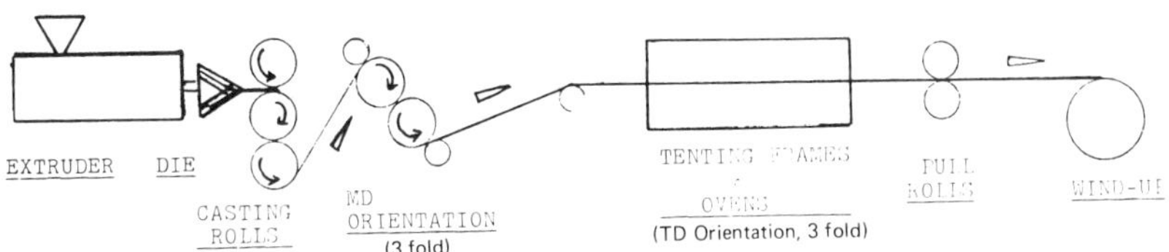

Figure 23. Biaxial-orientation of polystyrene. (Ref. 13.1.2 #17)

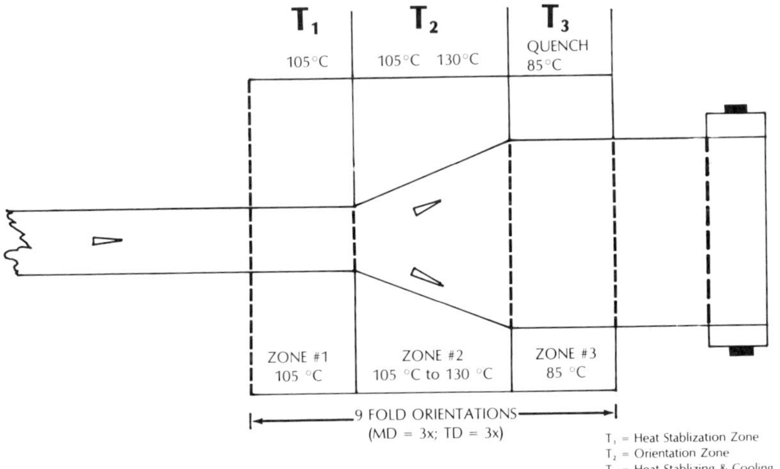

Figure 24. Biaxial orientation of polystyrene (schematic of tenting process). (Ref. 13.1.2 #17)

Because of the temperature control problems that are associated with this sequential stretching, many modifications of the tenter frame have been proposed. The most frequently suggested improvement is what might be called an accelerating tenter. Stretching in the machine direction is done simultaneously by draw rolls, which accelerate the film away from the nip rolls. Thus, orientation is introduced simultaneously into both directions of the sheet.

With the development of simultaneous biaxial tentering-type stretch units, it appears fair to state that none have the flexibility and degree of control that is possible with the two-step tentering process.

The major uses of oriented polystyrene films are for thin gage (0.5-3.0 mil) transparent windows in envelopes and boxes, and for simple overwraps for such items as tomato trays and Christmas wrapping papers. Its excellent dielectric properties are used to advantage in capacitors. It is also used for wrapping produce such as lettuce, cauliflower, and broccoli. In thicker gages, (3-20 mil), oriented polystyrene is used for sheet protectors, wallet inserts, and thermoformed articles of many types, e.g., cottage cheese tub lids, meat trays, tomato pack trays, cake domes, cookie trays, blisterpacks, etc. For such formed items, polystyrene provides the greatest toughness, clarity, and sparkle at the lowest cost. It competes easily with cellulose acetate and polyvinylchloride sheet in most of these areas.

Descriptions of the various process steps involved in the orientation of polystyrene are shown in Table 7 and Figure 25.

Table 7. Production of OPS

Process Steps
1. *Extrude* Sheet & *Quench*
2. *Reheat, Stretch,* (Machine Direction)
3. *Cool*
4. *Reheat, Stretch,* (Trans-Machine Direction)
5. *Cool*
6. *Trim* (Clip Waste)
7. *Wind-Up*

EXTRUSION → 10 mil sheet (Linear speed *a*/width *b*)

↓

QUENCH (CASTING ROLLS) → 10 mil sheet (Linear speed *a*/width *b*)

↓

PREHEAT SECTION (MD STRETCH) → 3 mil sheet (Linear speed 3*a*/width *b*)

↓

TENTER FRAME (TD STRETCH) → 1 mil sheet (Linear speed 3*a*/width 3*b*)

NOTE: 10 mil sheet can be used in tray forming for meat, candy, etc. The process for sheet extrusion consists of the first two steps. Film thickness, production rates and extruders are sized accordingly; normaly – 4½ or 6 inch extruders for heavy sheet; – 2½ to 4½ inch extruders for thin film. The ultimate decision will depend on specific markets, their location, and distribution costs.

Figure 25. Oriented polystyrene (schematic of producing thin OPS film).

POLYPROPYLENE (PP)

Ten producers in the United States account for 3.8 billion pounds of polypropylene per year. According to recent market predictions, polypropylene production will reach 5 billion pounds by the end of 1981.

Since propylene gas is a by-product of the cracking of crude oil during gasoline production and the production of ethylene from heavier naphthas, one could easily conclude that supplies will be adequate. However, in the early 1980's there is an excess of gasoline inventory and a switch to lighter oil fuel stocks for production of ethylene. This means that although supplies will be adequate one cannot expect any relief in price increases.

The continued strength of the U.S. dollar, however, will tend to support the prediction that exports will decrease and that prices will rise overseas. Overseas capacity for OPP is not sufficient to meet demand so it will take several years for European production (or new Near East facilities) to meet the demand. As noted in previous sections polypropylene supplies will be adequate but *not* abundant. There will be a supply problem in the mid 1980's when production is expected to reach 95 to 105% of *name plate* capacity.

Unoriented polypropylene film was introduced in the late 1950's. These films were cast from a sheet die onto a polished chill roll stack by standard methods. Unmodified cast film offers good yield, clarity, and grease and moisture resistance. When drawing or stretching cast PP, the properties are significantly changed or enhanced. Specifically, orienting or stretching improves impact resistance. It increases tensile strength (and the work function. Low temperature properties are enhanced, and optical properties, moisture barrier and grease and fat resistance are all increased.

There are two types of oriented polypropylene (OPP): biaxially oriented and uniaxially oriented (Figure 26). Heat stabilized OPP films that have been "stress relieved" support a wide variety of packaging applications. Unstabilized shrink PP is of lesser importance than the heat stabilized (stress relieved) OPP.

Oriented polypropylene (OPP) is made by either the blown film or the tenter frame process. The blown film process is better suited to produce films below 0.6 mil; both tenter frame and blown film processes can produce medium gauge OPP between 0.6 to 1.0 mil. However the tenter

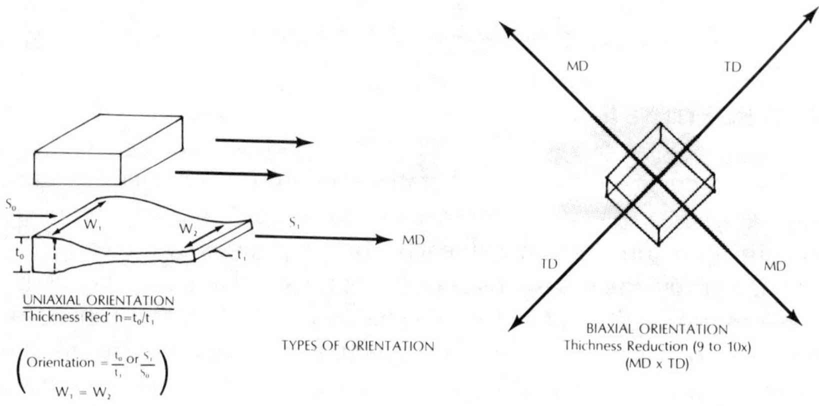

Figure 26. Uniaxial Orientation and Biaxial Orientation. Ref. 13.1.2 #1,3 – Plastic Films – J. H. Briston, Halsted Press, J. Wiley & Sons. Plastic Films in Packaging – C. R. Oswin, Halsted Press, J. Wiley & Sons. Source: Packaging Engineering and Modern Plastics.

frame has a clear advantage above 0.9 to 1.0 mil in the heavier gauges.

John Burke, product manager of OPP films for Mobil Chemical Company, predicted in early 1980 that OPP film sales could reach over 230 million pounds in 1981. This figure is over 2 times the 1977 sales figures for OPP – 100 million pounds. Optimism is based on the conviction that OPP is beginning to make impressive inroads into the glassine and foil packaging markets. Major applications are available in snack food packaging and bakery and cracker overwraps. The incentive is that opaque OPP is lower in cost than HDPE and metalized OPP is 5% less costly than PET or nylon. Many of these applications require films that are greater than 0.6 mil. This means large capital investments are required to go after this market. (75% of the OPP film market will continue to be in film thicknesses greater than 0.6 or 0.75 mil.)

An indication of corporate interest in the tenter frame process can be seen in the expansions of Hercules Inc., Mobil Chemical, Borden Chemical and Northern Petrochemical.

- *Mobil* Chemical started up a 50 million pound per year tenter frame in 1980 in Shawnee, Oklahoma.
- *Hercules* increased the capacity of its tenter frame operation to match its blown film capacity.
- *Borden* will have a 30 million pound capacity tenter frame in operation in 1981.
- And *Northern Petrochemical* will produce 20 million pounds by tenter frame in 1981-1982.

In all these cases emphasis is on a sparkling clear packaging film that is

12% to 20% less costly than cellophane. OPP films made by the tenter frame process exhibit 50% of the distortion of blown film at 275°F. These data indicate that the annealing or stress relieving action of the tenter frame is more efficient than the blown film process and produces a more stable film for use "as is" or as a laminate base.

This polymer can be converted into an oriented film which is either shrinkable or heat stabilized. Either type can be prepared through both the bubble or tenting processes.

Mono-axial Orientation – The film is first chill-cast then stretched at a temperature approaching T_m (120-140C). To improve clarity and to restrict spherulite growth one usually tries to stretch "in-line." Stretching can be done on sheet or bubble. Monoaxial orientation is done with sheet only. (Tubular films are collapsed, and opened before stretching). The film is heated over a series of rolls, which provide frictional grip, then stretched 8X to 12X its original length, the stretch temperature is such that the film is fully extended at this point. The temperature determines the amount of energy and strain.

Biaxial Orientation – In the bubble process, (similar to the Saran[TM] bubble process) a tube is extruded at about 200°C, rapidly quenched to approximately 10°C, then reheated to 150-160°C and expanded 2 to 10 times in diameter under air pressure while being axially accelerated 2 to 10 times by the take-off rolls at the nip (Figures 27-29). Film is heat shrinkable between 100-150°C (e.g. it will shrink up to 40-50% at 125°C). A heat-stabilized film with a shrinkage of less than 3-5% at 120°C can be obtained by reinflating the flattened bubble under lower air pressure and heating it for a few seconds near the melting temperature.

Biaxially oriented film can also be produced by a two stage process (Figures 29-31). The extruded sheet is quickly quenched and stretched in the machine direction over a series of slow and fast rolls then passed to a tenter where it is reheated and stretched sideways by mechanical extension (Figures 32 and 33). (Figure 20 also shows this approach). The advantages of this approach, compared to the tubular method, lie in the flatter film and in the ease with which one can anneal the film before it is wound. Usually the tenting process stretches a film 3X to 6X in each direction and anneals with a slight loss of 4% to 6%. Films produced by tenter are usually thinner, flatter, and have more uniform gauge control.

TM – Trademark Dow Chemical Co., Inc.

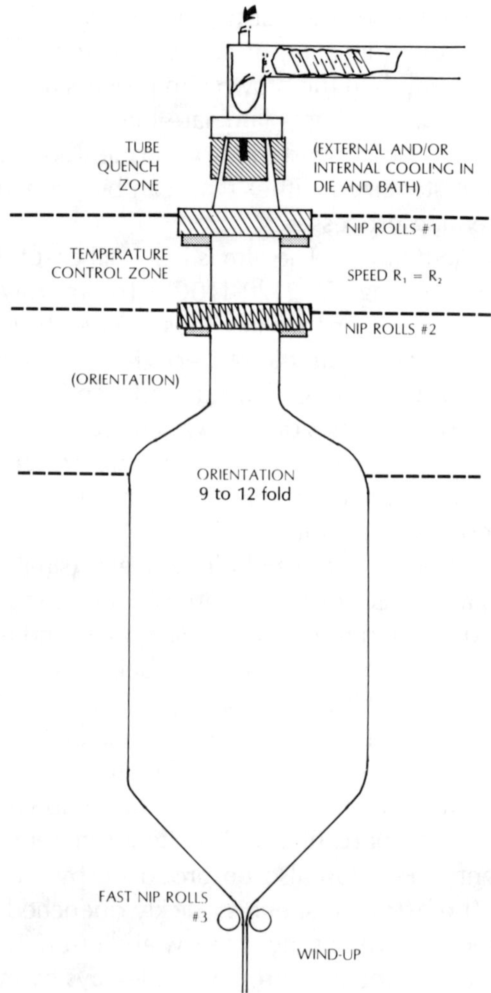

Figure 27. Orientation by the bubble process. Source: Modern Packaging Films, Edited by S. H. Pinnir Publ. Butterworth, London 1967.

A tenter can achieve almost 2 times the output of a conventional tubular process for a given film and gauge.

A summary of the properties of oriented polypropylene and the affect orientation has on polypropylene are illustrated in Tables 8 and 9.

In general, films oriented by the bubble process have more balanced properties in the machine and transverse directions than tentered films. This is illustrated in Table 10 (page 45) which compares properties

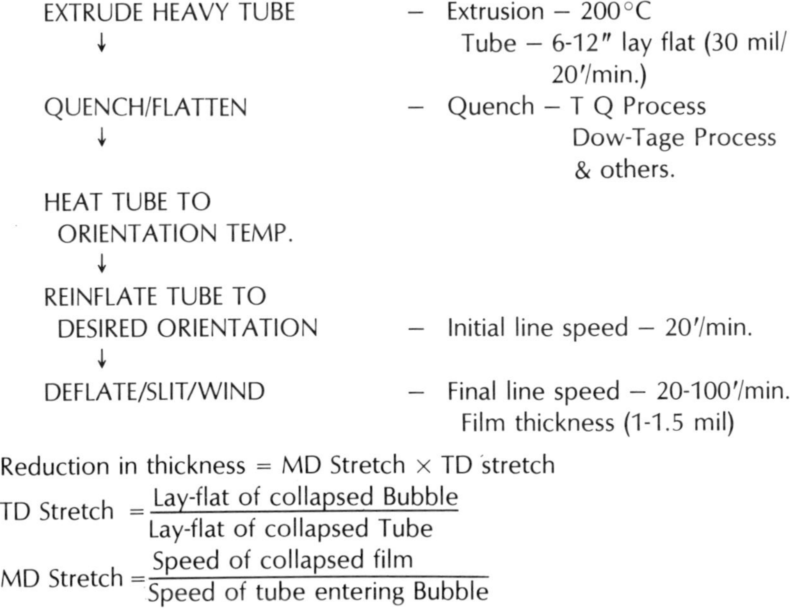

Figure 28. Schematic of OPP manufacturing process by trapped bubble.

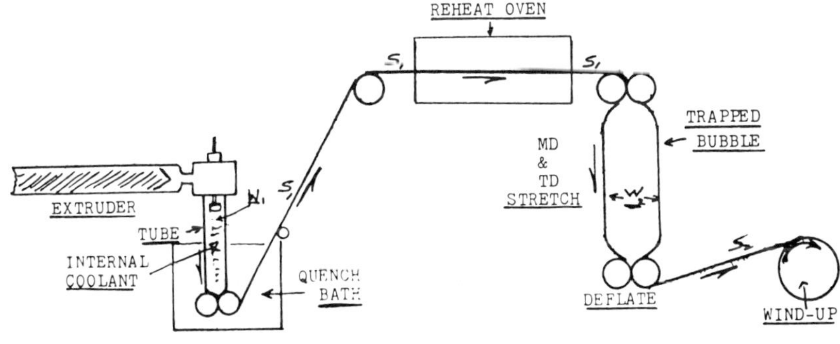

Figure 29. Orientation of polypropylene (trapped bubble process).

of representative polypropylene films oriented by these two methods. Theoretically, it is possible to achieve balanced orientation by tentering, but practically it appears to be more difficult to do this with crystallizable polymers than with noncrystallizable ones.

Unoriented polypropylene film is brittle at temperatures below 0°F and tends to shatter when used as an overwrap for frozen food; *biaxially*

Process Steps — 1. Extrude Tube, Quench
2. Reheat Tube
3. Inflate Tube (Biaxially Orient in Trapped Bubble
4. Collapse, Slit, Wind, etc.

Equipment — 3½″ Extruder - 6″ Tubing Die
— Take-off unit and quench bath.
— Blown film tower with feed rolls (R_1) heat unit and collapsing unit (R_2)

- *Tubing* — 10″ lay flat tube; 30 mil thick
- *Film* — 80″ wide 1→1.5 mil thick
- Orientation — 3x to 4x in MD and TD

Conditions —

$$\text{T.D.} = \frac{\text{Bubble Diameter}}{\text{Tube Diameter}}$$

$$\text{M.D.} = \frac{\text{Take-Up Roll Speed } (R_2)}{\text{Feed Roll Speed } (R_1)}$$

Figure 30. Production of OPP film. (Trapped bubble process).

Process Steps — 1. Extrude Sheet, Quench
2. Reheat, Stretch in Machine Direction
3. Cool,Split, Stretch in Transverse Direction
4. Slit, Wind etc.

Equipment — 1-6″ Extruder (300 HP)-48″ Sheet Die
— Roll Stand for Machine Direction Orientation (4x to 6x)
— 2 - Tenters (4x to 6x Transverse Orientation

Conditions — *Cast Sheet* — 30 mil, 48″ wide; 25′/min.
— *Orient Sheet* (MD) — 5 mil, 48″ wide; 125′/min
— *Orient Film* (TD) *2 Webs* - 1-1.5 mil; ~ 120″ wide; 125′/min

Figure 31. Production of OPP film. (Tenter Process - 1 extruder feeds 2 tenters).

Process Steps – (Same as Figure 30)

Equipment – 1 - 3½″ Extruder (50-60 HP)
– 1 - 24″ Sheeting Die
– 1 - Roll Stand for MD Stretch
– 1 - Tenter for TD Stretch

Conditions – 24″ sheet - 30 mil thick - 20′/min. Stretch in MD (4x to 5x)

Oriented Film – 120″ wide; 0.75 mil to 1.5 mil; 80 to 120′/min.

Figure 32. Production of OPP by single extruder/tenter.

Table 8. Properties of Polypropylene (Cast & Oriented)

	Cast P.P.	Oriented P.P.
Yield (in²/lb./mil)	31,000	31,000
Tensile Strength (psi)	6,000	25,000
Elongation at Break (%)	275	40-60
Elmendorf tear (gm/mil)	50-60	5-7
MVTR[1]	17-19	10-12
O_2 Tranmission[2]	3,700	2,800

[1]gm/m²/mil/ 24 hrs at 90% RH and 100°F
[2]c.c./m²/mil/24 hrs at 2 AM 0% RH
Source: Technical Data Bulletin and O. Sweeting "Science and Technology of Plastic Films, publ. J. Wiley Interscience pg. 286.

Table 9. Oriented Polypropylene Film Properties (Source - Ref. 13.1.2 Briston, Sweeting, Oswin et al)

Property	Copolymer (non-heat set)	Homopolymer (heat set)	Modified (heat set)
Tensile Strength (psi)	20,000	25,000	20,000
Elongation @ Break (%)	100	75	50
Secant Modulus, 1% (psi)	200,000	400,000	475,000
Yield, (in²/lb @ 1 mil)	30,000	30,000	30,000
Heat Seal Range (°F)	–	–	220-280
O_2 Transmission (cc/mil/m²/day/atm)	2,500	2,500	1,300
CO_2 Transmission (cc/mil/m²/day/atm)	8,000	8,000	37,00

Note: *Gloss* for all film above 90% @ 45°: *Haze* is less than 1%; *MVTR* is equivalent and depends on the "modification"
Source: Cryovac Exxon, Hercules, Northern Petrochem.

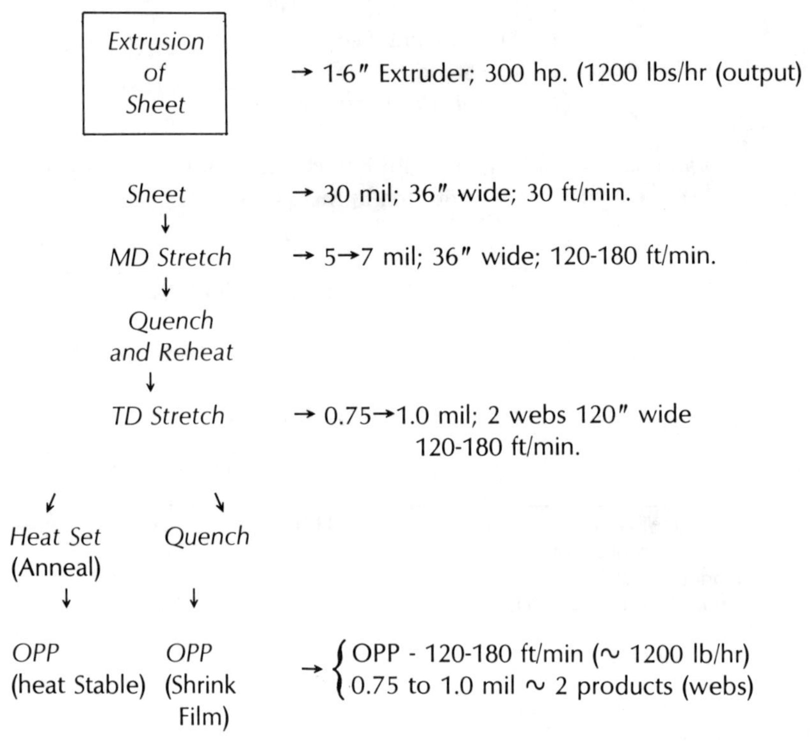

Figure 33. Schematic of OPP film lines (1 extruder; 2 tenters; producing both shrink and heat set OPP).

oriented film remains flexible at much lower temperatures. Oriented polypropylene film is used principally in wrapping such articles as phonograph records.

R. P. Hirt (Director of Packaging/Marketing, Hercules, Inc.) in a recent article in Modern Packaging noted that commercial biaxially oriented polypropylene films can be classified into two general categories and several sub-groups:

1. Heat Stabilized (non-sealable)
2. Heat Stabilized (heat-sealable)
    a) uncoated modified polymer types
    b) coated
    c) coextruded

Table 10. Properties of Oriented Polypropylene

Property	Bubble Process	Tenter Process
Thickness, mil	0.60	0.75
Tensile Strength, psi		
MD	26,000	17,000
TD	26,000	42,000
% Elongation		
MD	40-50	120
TD	40-45	20
% Shrinkage, 1 hr at 124°C		
MD	8	3.5
TD	15	3.5

Source: Cryovac Exxon, Hercules, Northern Petrochem.

The first category represents those films that are not heat sealable because they shrink upon application of heat and before sufficient "grab" or "seal strength" is developed to form an adequate seal. However, they are extremely important in combinations with other polymers or films. Examples of these composites or laminated films and their application and markets (i.e. laminates with cellophanes, Saran™ coated glassine, and Saran™ coated polypropylene for packaging greasy snack foods, and with paper as liners and innerlayers in multiwall bags) are discussed later in this section.

Thin gauge, non-sealable, oriented polypropylene film sells for 2¢ to 2.5¢ per 1000 in² in very thin gauges (approx. 0.45 mil).

There are three types of heat stabilized, heat sealable oriented polypropylene structures:

1. *Uncoated polymer modified base films* are used to replace cellophane as box overwraps (i.e. candy, food, drugs and cigarettes).

2. *Coated oriented polypropylene films* have heat activated coatings (fusable) which can be applied with conventional coating equipment from a lacquer or emulsion system. (They can also be made by coextrusion.) Film can also have adhesive layers, heat sealable coatings, coatings possessing special barrier properties, etc.

TM – Trademark Dow Chemical Co., Inc.

3. *Coextruded heat sealable films* are used in form and fill machines (i.e. bakery items, such as cakes, crackers, and pastries).

It has been noted that 100 million pounds of oriented polypropylene film used in 1977 was equivalent to two or three times that amount of cellophane.

The bright supply picture (assuming a stable international trade balance) and the excellent cost advantage that OPP has over its competition insures continued growth in sales and applications.

One dark cloud on the horizon we noted previously is the uncertain fluctuations of the dollar, the competition for resin domestically vs. export trade.

Resin Properties – The proper choice of resin is essential in optimizing the orientation process and final film properties. Basically there are two primary variables that affect processing and final film properties – the molecular weight and the polymer or copolymer structure (in the case of polypropylene, the comonomer is usually ethylene). In addition, additives contribute to variations in the orientation process and film properties.

Molecular Weight – Molecular weight is a primary variable that affects physical properties (i.e. tensile, elongation, impact resistance). Films prepared from high molecular weight polypropylene have good strength, impact, and orientability properties. In addition, as molecular weight increases, an increase in the melt strength of the resin can be observed. On the other hand, high melt strength or melt flow leads to more difficult processing and detracts from optical clarity. Films made of lower molecular weight material are characterized by being more brittle, less orientable, and having lower melt viscosity or strength. These lower molecular weight resins have good clarity and produce high extrusion or processing rates at equivalent energy input.

Melt flow or melt index values have been described in the literature as better suited to a specific process, as in the following examples.

1. *Tubular orientation* requires high molecular weight (M.I. 1 to 3) to provide adequate melt strength during extrusion and orientation. (Orientation improves the normally poor clarity of these resins.)

2. *Tubular processes* which include a quenching step do not require low melt index resin since the quenching stage in-

creases the melt flow at a rate sufficient to minimize distortion.

3. *Sheet extrusion* uses polymers in the 3 to 7 melt index range. (Melt strength is less critical than in the tubular process.)

4. *Sheet casting processes* can tolerate very low molecular weight resins (melt flows of 6 to 10) because melt strength is not required and good clarity is of prime importance.

Polymer Structure (Comonomer Level and Location within the Chain) – The second major variable affecting film and processing characteristics is comonomer content and structure. In polypropylene the comonomer that has been most effectively used is ethylene.

1. *Block copolymers* are not usually used as film resins. These materials contain large ethylene blocks (up to 15 to 20 wt. %) and yield non-transparent, elastoplastic type polymers.

2. *Random copolymers* usually contain just enough ethylene comonomer within the polymer chain to disrupt crystallinity (from 2.5% to 4% by weight of ethylene). This small amount of comonomer reduces melting point, improves flexibility and low temperature properties, and in addition produces films with excellent optical properties and good heat sealing characteristics.

Additives – As we noted previously, additives affect both extrusion and film properties. In addition, they must conform to present (and future) government regulations.

1. *Stabilizers* are used in all polypropylenes to reduce degradation during processing which in turn would produce film of lower molecular weight with its inherent poor film properties. Film grade resins usually contain low levels of non-volatile additives that do not plate out on the equipment during processing. (Stabilizers are also added that improve long term stability to heat and light, but in films for packaging these latter additives are not required.)

2. *Slip agents* are added to the film grade resin to overcome frictional forces encountered during processing. (The level depends on film thickness and orientation, processing parameters, and residence time at temperature.)

3. *Anti-block* agents are small particle-size fillers that are used to reduce film smoothness without detracting from film clarity. (This is usually accomplished by using fillers of specific particle dimensions and index of refraction).

4. *Anti-static agents* are occasionally used in films where long shelf life will be encountered and where dust accumulation is to be avoided. (These chemicals are essentially the same as those used in the injection molding resins). These additives are also used in resins designed to be converted into packages for packing powders.

Metallization – Metallized OPP adds another alternative to a market dominated by nylon and polyester. At first glance OPP should provide an economic advantage to nylon and polyester in flexible packaging. However, the properties of these three systems indicate that each will continue to have its own niche in flexible packaging.

In 1975 metallized polyester was quite common and its main use was in packaging coffee. Two years later metallized biaxially oriented nylon was being used in gas flush packaging and controlled atmosphere packaging, again coffee was the main marketing objective. Metallized oriented polypropylene followed in 1978.

Each one of these films has a place in the market. OPP (metallized) is lowest in cost, and the laminates using metallized OPP have the printing on the lower cost, non-metallized layer. For example, in the OPP/adhesive/Aluminum/OPP composite, the sealing layer is on the Aluminum/OPP side and the graphics are reverse printed on the inner surface of the exterior OPP layer. This means printing can be achieved on standard equipment and that an important cost advantage exists since printing scrap is generated on the low cost film not the metallized film. Metallized OPP offers excellent light protection and excellent high gloss.

Table 11. Barrier Properties of Metallized OPP Film (Ref. 13.1.1. P.Prince, J. Scharr Package Engr. Aug. '79 et al)

	Unmetallized	**Metallized**
Gauge (mil)	0.4	0.4
MVTR (g/sq.m./24 hr. at 25C and 75% R.H.)	2.2	0.6
Oxygen Perm. (c.c/mil/100in²/24 hrs/ atm at 0% RH	110	3.1

Source: Package Engineering August 1979. "Oriented Polypropylene–Metallization" P. Prince and J. Schaar, pg. 605-606.

OPP does not have sufficient oxygen barrier properties to be used for long shelf life items such as coffee, but many other markets are available (Table 11).

Table 12 shows that nylon has better oxygen barrier properties than PET or OPP (metallized or not). That is why laminates and coextruded film, bags, and shrouds contain nylon when used for shipping and packing fresh meat, and in controlled atmosphere or vacuum gas flushing of

Table 12. Barrier Properties of Metallized Laminates

	OPP[1] Laminate	Polyester[2] Laminate	Nylon[3] Laminate
MVTR (gm/100 in²/24 hrs) at 100°F 90% R.H.	0.1	0.1	0.2
Oxygen Permeability (c.c/100 in²/24 hrs) at Standard Conditions.[4]	3.0	0.08	0.05

[1]OPP/Printing/Adhesive/Aluminum/OPP (Sealing layer)
[2]Printing/PET/Aluminum/Adhesive/PE (Sealing layer)
[3]Printing/ON/Aluminum/Adhesive/PE (Sealing layer)
[4]Standard Conditions 70°F/50°R.H.
Source: Package Engineering August 1979 "Oriented Polypropylene–Metallization" P. Prince and J. Schaar, pg. 605-606.

large sections on sub-primal cuts which require extended distribution time.

Some typical costs were given in a recent Package Engineering article (8/79) by Prince (Inovpack) and Schorr (Schorr Industries).

Composition	**¢ per $1000 in²**
Polyester (0.5 mil)	4.2¢
Oriented Nylon (0.5 mil)	4.0¢
Oriented P.P. (0.5 mil)	2.0¢

Commercial Significance – There are five basic types of oriented polypropylene (OPP) films. The five film types include uncoated OPP films, modified uncoated OPP films, Saran™ coated OPP films, heat seal coated OPP, and acrylic coated OPP films. In addition to these, orientation can be balanced or unbalanced as required.

Background – Oriented polypropylene films have some inherent advantages over cellophane. For example, moisture protection of the five OPP film types is superior to nitro-cellulose coated cellophane film especially when the film is creased. It is also generally superior to Saran™ coated cellophane. The moisture barrier properties of OPP film is proportional to film thickness as one would expect. Dimensional stability is also a good indicator of film performance and directly affects the sales appeal of the final package. OPP films exhibit good dimensional stability during shipping and storage and over the temperature and humidity ranges normally encountered during retailing (Figure 34). Cellophane is not as stable under these conditions as OPP. Cost of OPP films is lower than cellophane on a cost/unit area or yield basis. The lower density of OPP

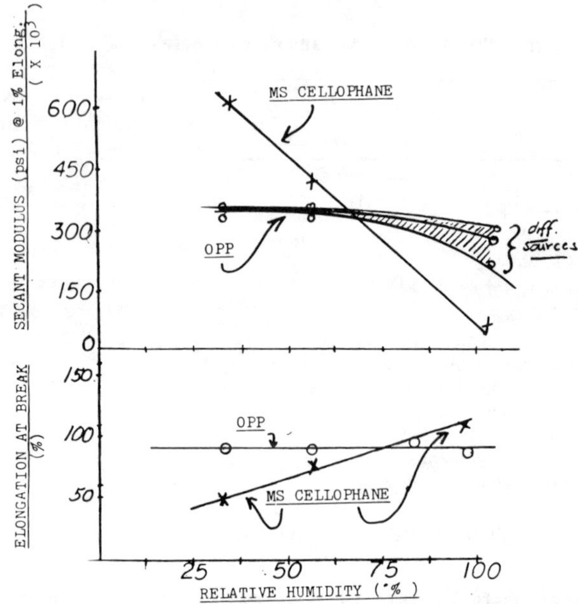

Figure 34. Effect of humidity on properties of cellophane and OPP. Sources: Package Engineering, Modern Packaging, Allied, UC&C.

gives it an attractive economic advantage (0.9 vs. 1.4 gm/cc). These are the basic reasons behind the current trend to replace cellophane with various OPP films.

Uncoated OPP Films – These can be used successfully in areas where product protection is not an important factor. In these application areas the film improves package appearance and provides a barrier to contamination by dust and foreign matter. Uncoated OPP films are relatively low in price. However, they have a restricted use in overwrap applications, their primary usage being a base for coated or laminated structures.

Modified Uncoated OPP – These films are made from blends of polypropylene and other polymers. These oriented "blends" come very close to the machinability properties of cellophane. The polymeric modifiers are structured to enhance the modulus of OPP and broaden the heat sealing range of the homopolymer below that normally encountered with unmodified resin. The increased stiffness allows the film to be converted on standard packaging machines that were designed for cellophane. These machines are usually overwrap machines used in the tobacco industry. The one drawback, weaker seals, is not considered to be of major consequence since the final seal is equivalent to the seals encountered in cellophane overwrap applications.

Simple addition of static eliminators, accurate temperature controllers, and air assist manifolds can easily convert a machine to handle modified uncoated OPP, in a high speed film application.

Saran™ Coated OPP Films – Saran™ coated films have found wide acceptance where an oxygen barrier is required. The relatively high price and poor machinability of Saran™ coated OPP has made it a poor competitor for Saran™ coated cellophane. In general, Saran™ coated cellophane performs better on packaging machinery than the comparable OPP. As the coating process improves, the Saran™ coated OPP is making inroads into the cellophane overwrap market. The environmental pressures related to the making and converting of cellophane film will continue to be a deterrent to the economic advantages of cellophane.

Heat Seal Coated OPP – The strong, clear, high modulus, oriented homopolymer films approach the properties of cellophane. Heat sealing limitations are normally overcome with heat sealable coatings when used in packaging snack foods, candy, and pet foods. Typical heat seal layers include polyethylene (of low molecular weight), ethylene vinyl acetate, polypropylene copolymers, and in some cases coated cellophane. An additional application which does not require heat sealing is a grease barrier in paper laminates. These glossy scuff resistant coatings or composites are also used in book covers and record jackets.

Acrylic Coated OPP Fioms – The properties of these films are similar to those of coated cellophane. Their properties were engineered to run on the same type of overwrap machines at equivalent performance. The typical properties of acrylic-coated films are shown in Table 13. (A general comparision of shrink films is discussed in the Properties section). These data have been presented by various authors and Appendix 13.1.1 lists many of these sources.

The data indicate that the heat seal range of acrylic coated OPP is not as broad as that of cellophane, but it is broad enough to provide good sealing properties without modification to the sealing equipment normally used for cellophane.

Markets for OPP Laminates and Products – Polypropylene film stands out as the major participant in the future development of flexible film packaging applications. Oriented polypropylene film, coated, laminated, and "as is," promises to have a continuing impact.

Cellophane replacement has been a major target for polypropylene films. Several product categories such as tobacco have converted almost entirely to OPP. At times this has meant a better package but at a lower cost. Snack foods is another area where oriented polypropylene lamin-

Table 13. Properties of Acrylic-Coated Biax-OPP Film

Property Mechanical	Test Method	Acrylic Coated OPP — Values
Tensile strength (psi)	ASTM-D-882	20,000
Impact strength	TMI Impact Tester	16 kg/cm./mil
Stiffness	Handle-O-Meter	3 gm./inch width
Tear strength	ASTM-D-1922	5.2 gm.
Heat Seal Range	¼ psi./2 sec.	210-290 F (100-140 C)
Coefficient of friction	ASTM-D-1894	0.25
Optical Properties		
Gloss	ASTM-D-2457	87 @ 45 degrees
Haze	ASTM-D-1003	1.0%
Barrier Properties		
WVTR	ASTM-E-9666	0.45 gm/100 sq.in./24 hr.
Oxygen	ASTM-D-1434	130 c.c./100 sq.in./24 hr./atm.
Stability Parameters		
Shrinkage	280-290°F (max. use temp.)	10%
Humidity	No effect.	

ates are making in-roads. One such material is Mobil Chemical's acrylic coated OPP laminated to a coextruded and oriented polypropylene, and ethylene-propylene copolymer combinations.

A Hercules composite has an OPP core with sealable coats on each side. This is finding application in "cookie trays," according to recent studies and articles in Package Engineering.

A new area where it is expected OPP will become more involved is in the metallized film area. The current pouch is metallized polyester with a LDPE sealing layer. With a nitrogen flush the coffee pouch (for example Elkin's) has at least a 90 day shelf life.

Studies at some food institutes have indicated that metallized films are increasing in volume and that they have a good potential in the area of fatty foods, such as, potato chips. Metallized polyester, nylon, and polypropylene are currently the major composites. Oriented polypropylene (coated and/or laminated) is believed to be the major contender in the next five to ten years, especially with polypropylene costs being very attractive relative to the others.

The retortable pouch for "ready-to-eat" items is maturing in the marketplace and at the present time there are no clear cut applications for heat set OPP in current applications. Many compositions are being tested but none is an obvious major contender. OPP could replace HDPE in the PET/Foil/HDPE (or OPP) laminate which has a sealing range of 250° to 300°F. This laminate has good low and high temperature performance. It's typical use could be in the retortable pouch.

Some OPP combinations used in ths snack food area are:

OPP/Glassine/PVDC

This laminate contains a "reverse printed" OPP layer and a heat seal coat of PVDC. Sealing range is reported to be 215-280°F. The laminate has good moisture barrier properties and is used for snack foods.

OPP/OPP/PVDC

This is a coated laminate that seals between 220-290°F with graphics reverse printed on the OPP. This composition is also used in snack foods. According to Package Engineering the laminate has good *hot tack* (outer OPP layer is modified for slip and the inner OPP is PVDC coated for additional barrier).

OPP/Metallized/OPP/OPP Copolymer

This combination is a coextruded OPP/OPP copolymer combination laminated to metallized OPP. (The metallized layer is in the sandwich construction to protect the coating and give additional barrier.) It has relatively high heat seal range, 255-290°F and has excellent light, gas, and moisture barrier properties. This is also sold to the snack markets.

OPP/PVDC Adhesive/OPP/Ionomer

This is a coated laminate with a heat sealable ionomer layer as an inside coat. The outer layer of OPP is a slip modified material. The PVDC adhesive provides integrity as well as gas and moisture barrier properties. Finally, the ionomer is coated on the OPP as the heat seal coat. This laminate is finding application in gas flush, snack foods, and nuts.

The broad versatility of biaxially oriented polypropylene can be illustrated by listing a typical polymer series and their application areas.

> *Heat stabilized OPP films* are used in lamination with cellophane, glassine, PE, or PVDC-coated OPP film. Applications include snack, candy, bakery, coffee, cheese, and meat packaging.
>
> *Modified OPP for heat sealabilty* is used for printed overwraps in candy, baked goods, pet foods, tape cassettes, and personal products.
>
> *OPP with Saran™ coating* is used in lamination with PP, PE, and cellophane. Applications in snack foods, bakery, candy, and cheese wraps.
>
> *OPP film with PVDC coating* has low oxygen permeability for gas flush applications in meat, cheese, or coffee packaging applications.

OPP used as an overwrap an horizonatal form/fill/seal machines, used for bakery, candy and commissary items.

Future – With a growing variety of products and applications OPP film will continue to expand, as it has in the past ten years, especially with the lower cost relative to the alternatives. In 1974 predictions were made (and have been substantiated) that the market would expand 10 to 12% a year until 1980, after which it is expected to level off at 8 to 10%.

Two applications stand out as having particular attraction as we noted previously – *cigarette pack overwrap* (40 million pound per year potential) and *retortable pouches* (polypropylene/aluminum foil/polyester constructions). The polypropylene gives food contact and a heat seal surface. The aluminum foil is the barrier and polyester supplies impact and scuff resistance. Recent articles in many packaging magazines have stated that the retortable pouch may be emerging in the 80's. Whether polypropylene will become a major shareholder in this market, remains to be seen. Certainly, the economics of the film as a laminating base are attractive.

The rapidly changing relationships of monomer costs will favor polypropylene for the foreseeable future.

A definite trend emerges when one reviews the published data about biaxially oriented polypropylene film in the past seventeen years. For example:

1965 . . . . OPP had 3 to 4% of cellophane market.
1970 . . . . Penetration reached 14%.
1974 . . . . 28% of market was captured by OPP with *no* major overall growth in the market.
1976 . . . . 40%.
1978 . . . . Penetration of the cellophane market reached 63%.
1980 . . . . Penetration of the cellophane market reached 70%-73%.

In the U.S. it is predicted that polypropylene will probably capture a maximum of 90% of those markets previously dominated by cellophane. (In Europe penetration is about 40 to 45% and may level off at 55-60%.)

In Japan polypropylene has taken the market because the film is manufactured internally and cellophane is imported (there is no significant domestic production of cellophane).

At the present time cellophane producers are setting their prices low to forestall additional penetration of OPP, but with plants operating at

60 to 70% capacity the battle for the market will be long and drawn out.

The key is price per unit protection.

According to the literature, film cost for uncoated OPP has risen from 2.3¢ per 1000 sq. in. in 1973 to 2.7¢ per 1000 sq. in. in 1978. Coextruded OPP has risen from 2.6¢ per 1000 sq. in. in 1973 to 3.2 to 3.3¢ per 1000 sq. in. in 1979. PVDC coated cellophane (and perhaps this is an unfair comparison..CJB) has risen from 4¢ per 1000 sq. in. to 7¢ per 1000 sq. in.

Caution should be taken not to make comparisons between"apples and oranges" so to speak. It's important to compare film on the basis of the degree of protection it supplies per given area.

POLYETHYLENE TEREPHTHALATE (PET)

Market Overview – Polyester film (PET) is made by the condensation reaction of terephthalic acid and ethylene glycol. Orienting the film produces balanced properties. The film has a temperature range of –80°F to as high as 400°F and is dimensionally stable at various levels of humidity. PET films exhibit good resistance to oils, organic solvents and chemicals – strong alkali attacks PET.

The different forms or types of PET films commonly used in packaging are:

1. *Standard (uncoated-oriented).* Treated on one side to improve adhesion of inks, adhesives, etc.
2. *Coated one-side* with Saran™ for use in laminated films.
3. *Coated two-side* with Saran™ (for high barrier) with coating engineered for heat sealing other PVDC coated substrates.
4. *Thermoformable films (sheet).* Uncoated or coated for shallow containers or trays.
5. *Heat shrinkable.*

Most of the compositions under discussion are converted, laminated or extrusion coated, with the type of coatings noted above.

Processed meat, cheese, and snack foods are the major food packaging applications. Applications where controlled atmosphere packaging or inert gas packages are required, are fruitful markets for PET composites.

Saran™ coating on PET provides the oxygen barrier necessary to prevent light-induced loss of color to cured meat.

Mr. Reiter of DuPont has written an excellent review summary of PET films in Modern Packaging (12/78). He notes that coatings have specific functions, for example,

Saran™ – to inhibit mold and oxidative degradation.

Polyethylene – To give hermetic seal integrity. Some applications of this versatile material include:

PE coated (1½ to 2 mil) for sealing film for boil-in-bag flat pouches.

Heat shrinkable (not heat set) for bag and tubing for poultry,hams, and sausages.

As an overwrap for cigarettes, bakery goods, snacks, toys, paper products, hardware, clothes, etc.

Some common polyester films include MYLAR™ (DuPont), SCOPAK™, SCOTPAR™ (3M Co.), and CELANAR™ (Celanese). Films are available in widths up to 58″ and 0.2 to 10 mils.

Technology and Properties – The commercial value of orientation is well illustrated by this polymer, which is made into significant quantities of oriented fiber, heat-shrinkable film, and heat stabilized oriented film. There are no applications for the material in its unoriented form because; if crystalline, it is extremely brittle and opaque, and if amorphous, it is clear but not very tough. When biaxially oriented, however, film acquires exceptionally good properties (Table 14).

Table 14. Properties of Oriented Polyethylene terephthalate

	Machine Direction	Transverse Direction	Amorphous & Unoriented
Tensile Strength, (psi)	19,000	18,000	8,000
Tear strength, (g/mil)	500	500	20
Shrinkage (30 min at 150°C) (%)	+5.0	−2.9	0
Elongation (%)	16.5	188	450

Table 15 lists various amounts of shrinkage which can be obtained with polyester films. These are all clear films and of the same chemical composition. The differing properties reflect differences in heat-stabilization (crystalline content) and in the amount of orientation in the two directions.

The special properties of polyester film – for example, it's strength, clarity, and good dimensional stability – make it an ideal base for packaging laminates.

Polyethylene terephthalate films are formed by extruding the polymer through a sheet die and biaxially stretching the sheet under controlled conditions into a strong, flexible, clear web. The oriented film is heat set. This means that internal stresses or "shrink forces" are reduced or eliminated. Subsequent coating operations are performed if one desires

Table 15. Shrink Properties of Polyester Films (Ref. 13.1.1, No. 34)

Film	Temp. (°C)	Shrinkage, 15 min (%) MD	TD
Stabilized Polyester	100	<1	<1
	150	2-3	2-3
Shrinkable Polyester I	100	50	20
Shrinkabke Polyester II	70	2.5	2
	80	21	22
	100	30	35

special barrier properties, slip characteristics, or heat sealability.

Typical properties of a 0.50 mil oriented uncoated PET heat set film are listed in Table 16.

Table 16. Properties of Uncoated PET Oriented Film

Thickness (mils)	0.50
Yield (sq. in./lb.)	42,000
Density (gm./c.c.)	1.4
Ultimate tensile (MD/TD) (stress at yield) (psi)	26,000/30,000
Elongation @ Break (MD/TD) (%)	125/90
Modulus (psi)	550,000
Haze (%)	1 to 5
Shrink @ 150C/30 min. (%)	1 to 2
W V T R (gm/100 sq.in/24 hrs.)	2.5 to 3
Oxygen Permeability (c.c./100 sq. in./ 24 hrs.)	8.5 to 9.1
Thermal Properties	
Melting Point(°C)	245-255
Processing Temp. (°C)	195-210
Max. Service Temp. (°C) (continuous)	150

Uncoated polyester films used in packaging applications are used as a single web in a lamination and can be reverse printed or metallized. One sided PVDC coated polyester is used in cheese and processed meat. Formable varieties can be drawn into shallow trays with another laminate, and the non-formable laminates are used in bags or for lids.

Two sided PVDC coated films are used as a general overwrap and as a base for laminating additional heat seal layers.

Heat shrinkable PET is converted into bags and tubes for smoked meat and spreads.

A study by Rutgers University scientists, in conjunction with ICI America, relates protection of fatty foods or shelf life to degree of metallization of the film. Elkins™ Coffee container is a good example. It consists of a metallized polyester, coated with a low density polyethylene sealing layer. (Note: the LDPE protects the metallized layer and provides a heat seal coating.) This coffee pouch keeps the coffee fresh for a minimum of 90 days.

Other composites are listed below. (Many of these have the heat seal layer of PE, or PP. Barrier properties are obtained with a PVDC barrier coating or use PVDC as a barrier adhesive, with metallization.)

ORIENTED, UNCOATED PET is used as a base for laminates, snack foods, retort pouches, medical supplies, boil-in-bag, hardware, gas flushed coffee, and frozen foods. It can be metallized, coated with barrier adhesives and usually possesses an inner "heat seal" layer or laminate (MDPE, LDPE, or PP).

ORIENTED, UNCOATED, FORMABLE PET is used in cook-in-bag applications or in trays.

ORIENTED, UNCOATED, TACK HEAT SEALABLE PET is used in bake-in-bag applications (bread, rolls, buns) brown and serve.

ORIENTED, UNCOATED, HEAT SHRINKABLE PET is used in frozen poultry bags and for frozen processed meat.

(One item should be noted: centrallized packing of frozen turkeys and large birds has been dominated by radiation crosslinked polyethylene made and sold by Cryovac™ and will continue).

Structures of the uncoated PET laminates – for example, laminates with LDPE, MDPE, HDPE, LDPE/EVA or foil/OPP – are noted for toughness, good converting performance on machines, temperature stability (both low and high) and excellent gloss and clarity.

ORIENTED COATED PET is used in laminates and have the following structure and uses:

ORIENTED, COATED

Laminated Films

1 side PVDC coated PET/LDPE

Pouches for processed meat

Pouches for cheese

Pouches and bags for coffee and oily solids

Laminated Films for Formed Trays

1 side PVDC formable Sheet – PET/LDPE
Processed meat tray

Laminated, Coated Structures for Snack Foods and Candy

2 sided PVDC-PET/uncoated OPP
2 sided PVDC-PET/coated OPP
2 sided PVDC-PET/coextruded polyolefins
2 sided PVDC-PET/PVDC coated cellophane
2 sided PVDC-PET/LDPE

A property comparison is listed in Table 17.

Table 17. Property Comparison of Competitive Polyester, Cellophane and OPP Packaging Films*

	Thickness (mils)	Stress @ Yield (MD/TD-psi)	Elongation @ Yield (%)	MVTR (1)	Oxygen Permeability (2)
Polyester					
UNCOATED	0.5	25,000/30,000	125/90	2.8	9.0
COATED					
1-side PVDC	0.55	23,000/28,000	120.90	0.9	0.4
2-side PVDC	0.6	23,000/28,000	110/85	0.5	0.3
Cellophane					
BARRIFR COATED					
A	2.5	18,000/8,000	20/50	0.4	1.0
B	2	18,000/8,000	22/50	0.4	1.0
Oriented PP					
O.P.P.	0.5	20,000/15,000	70/120	0.4-0.6	80
PVDC COATED	0.7	20,000/28,000	90/130	0.2	85
ACRYLIC COATED	0.7	8,000/24,000	365/60	0.4	1.3

(1) gm/100 sq. in./24 hrs.
(2) cc/100 sq. in./24 hrs./atm.
*Abstracted from "Polyester Films for Packaging Applications" by J. J. Ericson – SPE ANTEC May 8, 1979 Preprints .

It's understable why fast changing economics are making PET more attractive.

Several factors, however, tend to favor OPP. These are:

1. Ease of processing
2. Cost of Raw Material
3. Environmental pressure on production of cellophane
4. The rising cost of aromatic petrochemicals that are essential to the PET.

Polyethylene terephthalate film, because of its inherent toughness, is utilized in such applications as photographic film base, special drafting paper, and laminates such as the boil-in-bag type. This laminate, 0.5-1.0 mil polyethylene terephthalate film and 1-3 mil polyethylene, when sealed into a pouch for frozen vegetables, remains flexible at subzero temperatures, is tough, and does not shrink when dropped into boiling water.

Finely slit vacuum-metallized polyethylene terephthalate film is used in simulated metallic yarns. Decorative pressure-sensitive decals and communication satellite balloons (Echo I and II) are also made from metallized polyethylene terephthalate film.

VINYL CHLORIDE POLYMERS AND COPOLYMERS

Overview – PVC is a very versatile polymer. It fits a wide variety of applications and food packaging.

Vinyl resin can be converted into film by standard methods of blown film extrusion of plasticized vinyl and vinyl copolymers, flat die cast extrusion, solvent casting (vary small capacity), and calendering. These films range from the very rigid and hard to the soft and pliable stretch films. PVC film can be converted into shrink film by "freezing in" the stresses during orientation or into stretch film by "stress relieving" by heating under constraint after drawing into thin gauge material.

The largest demand for PVC packaging film is in-store wrapping of fresh meat, and lately in centralized packaging of poultry and poultry parts, vegetables, and produce at the source. PVC films possess excellent properties in strength, permeability, and retention of strength under tension. In 1978 it was estimated that 120 to 125 million pounds of PVC film went into the packaging field. Blown film dominates the market and solvent cast films are a very minor factor.

Lately, high-speed overwrap machines have utilized stretch PVC films, because they produce tight packages with good heat seals.

A comparision of PVC with a new EVA pallet wrap being marketed by DuPont shows that at comparable properties and weight there could be good competition brewing between PVC and EVA copolymers.

For example,

PVC (0.8 mil) has a Spencer puncture resistance of 8 in. lbs/mil.

EVA-3135 (1.0 mil) has a Spencer puncture resistance of 12 in. lbs./mil.

Tensile strength, elongation, wrap force, stretch, and weight of film (per standard pallet load) are all at a stand-off. This means the final question to be answered is what is the best performance one obtains per unit? (lbs./load, or sq. in/load).

Azotherm, a Cincinnati based company, is marketing a new line of heat sealable uniaxially oriented PVC films. The films have excellent clarity, high tensile strength, good dimensional stability, and gloss. These films are used in laminations for packaging snack foods.

At the present time food service operations use plasticized PVC stretch film, typically in such places as restaurants, schools, and hospitals.

A recent development, an opaque PVC film containing low plasticizer levels, is being used as a paper replacement in delicatessens. The white PVC film is a better moisture barrier than paper and keeps sliced processed meat fresher. (I'm sure the reader is comparing high density polyethylene (HDPE) that can be converted into an opaque moisture barrier approaching the appearance of glassine). HDPE is also a contender for these applications.

To fully evaluate the cost of PVC vs. its competitors (for example, EVA) as a bundling film or pallet wrap, one must compare both films at maximum stretch (without loss of "holding power" or tension) on a standard container or pallet load to determine the weight of film/load required for protection. This is not an easy approach but it will give meaningful data to the customer and an excellent tool for the Director of packaging.

PVC film is selling (1980-81) at 3.3 to 3.7¢/1000 in^2 per mil thickness.

Processes and Applications – Biaxially oriented films of polyvinylchloride and their copolymers with vinyl acetate are being used extensively for both shrink and stretch wrapping. Most of the commercial films are produced by the extrusion process followed by stretching by either the bubble (i.e. blown film) method or by the tentering process (two stage). Biaxially oriented PVC has three major advantages over its unoriented state. Increased impact strength, improved thermal stability (i.e. improved low temperature properties), and improved optical properties.

Extrusion processes have not always been the preferred method of producing PVC shrink film. Reynolds has been one of the largest producers of PVC oriented films. One of their processes, according to USP #3,036,147 (Table 18) was based on solvent casting a tetrahydrofuran solution of PVC resin. According to this patent the PVC is dissolved in the

Table 18. Schematic of Oriented Solution Cast PVC Film Production

STEP 1 – MIX PVC, THF, ESTERS (100, 300, 30)
STEP 2 – FORM PVC SOLUTION (160-180°F)
STEP 3 – FILTER
STEP 4 – CAST & DRY (180-225°F)
(4 mil film)
STEP 5 – STRETCH (180-225°F) AND ORIENT $9\times(3^2)$
STEP 6– THIN - FILM (0.5 mil)

(Thin, oriented PVC, plasticized with non-migrating, epoxidized plasticizer.) See USP 3,039,147 to Reynolds.

suitable solvent and cast on an endless steel belt. The dried film is stripped off the belt. The uniform flat sheet is then oriented by a standard two-stage tentering process to produce a ½ mil heat-shrinkable film. Since low temperatures are used throughout the process, virtually no polymer degradation takes place, and the scrap polymer can be completely recycled.

The Film Division of Cadillac Plastic & Chemical Co. also marketed shrinkable solvent-cast PVC film in the late 1960's Cadco™ S-401. The film had wide heat-sealing range (started at 305°F). Shrinkage starts at 250°F. The general purpose film had excellent clarity and was FDA approved.

This sequence of dissolving, filtering, casting, drying, and solvent recovery requires sophisticated equipment to satisfy the environmental constraints, with the likelihood of excessively high operating costs and poor productivity.

One should note that the thin oriented PVC films made by either the Reynolds patent or an alternative drum casting process outlined in USP #3,050,784 (Figure 35), whereby a PVC plastisol is cast on an endless mylar sheet to form a ½ mil film, can be used very effectively for in-store packaging of fresh meat and produce. This is essentially a stretch film application where the thin gauge and exceptional clarity and sparkle add an exceptional aesthetic quality to the product.

However, as noted before, the trend has been away from solvent casting for some time and these processes eventually will be of academic interest rather than commercial.

Most of the plasticized PVC stretch film produced in the United States involves the process steps outlined in Figure 36. A slightly modified blown film extrusion line can produce acceptable PVC stretch film at rela-

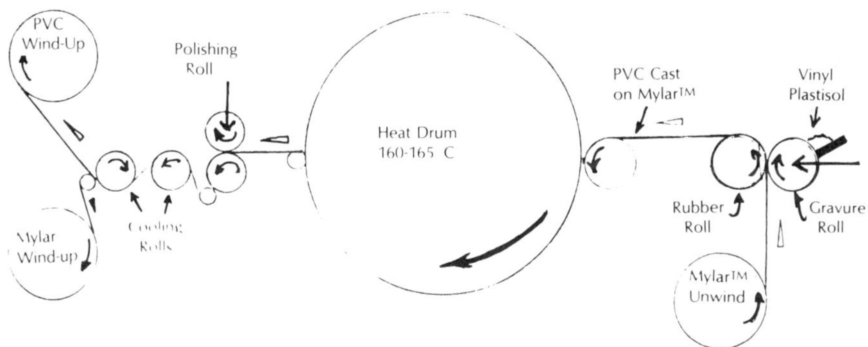

Figure 35. Process for making thin PVC stretch film (0.5 to 1.5 mil). Process involves casting a PVC Plastisol onto a temperature resistant film (i.e. Mylar™); gelling the composite; polishing the surface of the PVC; and delaminating and winding up the individual films. USP 3,050,784.

EXTRUDE – (circular die)
↓
Blown Film for Stretch Film Application

STANDARD FORMULATION

PVC	–	100
DEHA	–	50
DOTM	–	2
Stearic Acid	–	0.5

Figure 36. PVC stretch film process.

tively low operating costs. The properties of PVC stretch films used in meat, produce, and as an overwrap are compared in Figure 37. Thin gauge, PVC stretch films stress relax with time. This means that although PVC stretch film is ideal for in-store packaging of perishables, long term heavier duty applications may require heavy gauge, i.e. 3 to 6 mil. In these cases it may make better economical sense to use the polyethylene copolymer stretch films with their higher yield values.

Oriented PVC films for shrink packaging or heat stabilized for laminating base stock are produced on a standard sheet extrusion line followed by two stage orientation as outlined in Figure 38 (it essentially

APPLICATION	OVERWRAP	PRODUCE	MEAT
Thickness (in)	0.001	0.0008	0.0008
Yield (in^2/lb)	21,400	28,600	28,600
Tensile strength (psi)	8,000	4,000	3,000
Elongation @ Break (%)	120	150	150
Resistance to (moisture)[1]	16	10	6.5
Resistance of (Oxygen)[2]	63.0	10	4

Note 1. gm/24hrs/100 in^2-24 hrs. 90% RH. - 100°F
2. C.C./Mil/100 in^2-24 hrs./ATM 0% RH.
*Based on data from Comm'l Films.
(Reynolds and Others)
Source: Plastics Films and Packaging by C.R. Oswin, Halsted Press, Publ. pg. 68.

Figure 37. Properties of plasticized PVC stretch film. (Ref. 13.1.2 No. 1-3)*

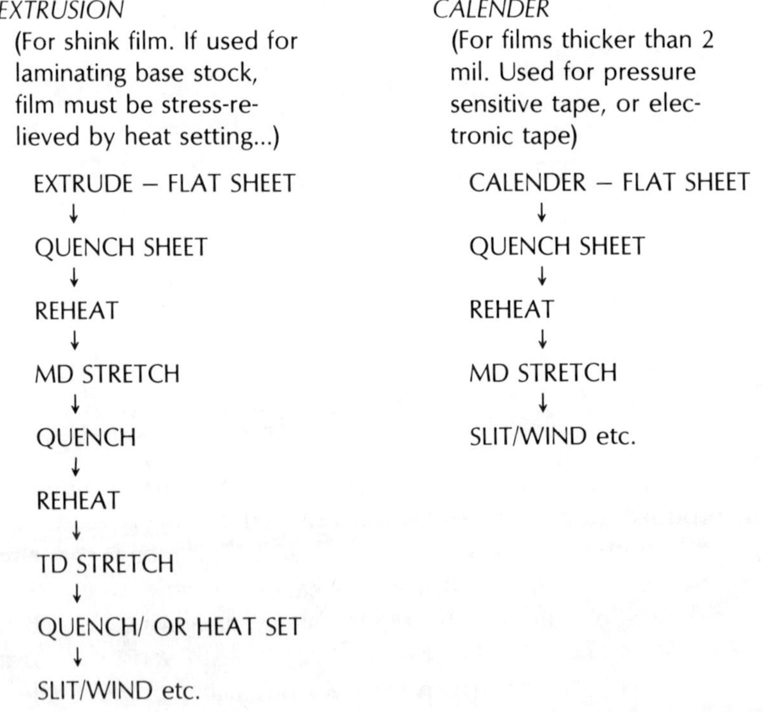

Figure 38. Processes for producing PVC oriented film.

involves the same process steps as those described in Figure 32 and Figure 33).

It is quite obvious that given the proper film parameters, the extrusion approach (whether standard sheet followed by two-stage orientation, or blown film followed by two-stage orientation) is preferred as being less costly and more productive.

Rigid PVC shrink film is produced by similar processes. For example there are two alternatives (Figure 39). Vertical or horizontal "straight" tube extruders can be used to produce a "seamless" tube (4 to 6 mil) which is slit and stretched on a conventional two-stage tenter frame, or rigid PVC film can be produced by calendering following by similar two-stage tentering.

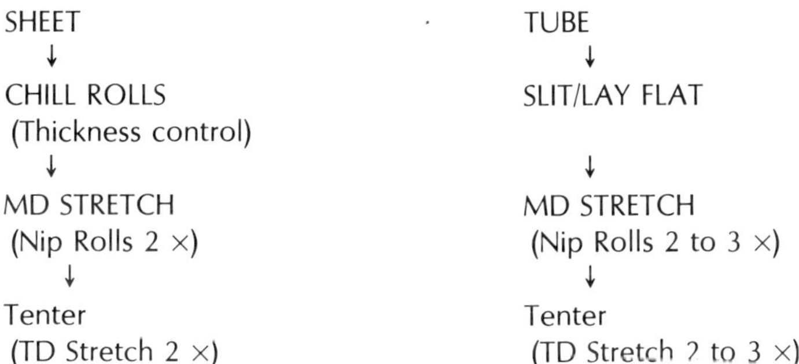

Figure 39. Processes for producing oriented PVC film (OPVC).

High quality rigid PVC shrink film is marketed in Europe very effectively. Its properties are compared with a "plasticized" version in Table 19.

Oriented vinyl films are used principally as stretch or shrink wrappings for meat and poultry, skin packaging of display cards, and for convenience. Oriented PVC films used in shrink applications are used in wrapping produce (tomatoes, displays in trays). Some very new developments are emerging which produce PVC outdoor siding from heavy gauge oriented sheet. Twenty year outdoor stability is being claimed by Solvay™.

Methods of Film and Sheet Manufacture – Plastic World had an interesting article in their July 1979 issue. Technology trends in PVC calen-

Table 19. Properties of "Rigid" Oriented PVC Film[1]

Property	Unplasticized PVC	Oriented Unplasticized PVC
Thickness (in)	0.0012	0.001
Yield (in^2/lb.)	17,100	20,100
Ultimate Tensile Strength (psi)	19,400	24,000
Elongation at Break (%)	50%	20%
M.V.T.R.[2]	32	16
Oxygen Permeability[3]	52	35

[1]Film made by calender followed by tentering.
[2]gm/mil/100 in^2 – 24 hrs. at 90% RH·100°F
[3]cc/mil/100 in^2/atm. 24 hrs. at 0% RH
Source: C.R. Oswin "Plastic Films and Packaging" Halsted Press pg. 65.

dering were described and evaluated in relation to costs, capital expenditures, and productivity. As noted in the article, there are 155 production lines installed in the U.S. to produce calendered PVC. The thinnest gauge that one can obtain with calenders is 2 mil vs. the blown film process capability of ½ mil to 1 mil. Although calendered film is not a factor in stretch or shrink applications economics could be affected. For example compare the output with invested capital in Table 20, and then realize that a calender produces very uniform film capable of being used as a basis for composite laminated film and to feed *several* oriented film converters.

Table 20. Methods of PVC (Sheet & Film) Manufacture (Ref. Plastics World - July '79)

	Calendar	Extruder Calendar	Extruder (Brown Film) (4½")	Extruder (Sheet) (4½")	Plastisol (Cast)
Machine Cost ($ Million)	1.2-3.6	1.2-2.9	0.5-0.8	0.5-0.8	0.5-0.9
Rate lb/hr.	800-6500	600-1600	–700	750	800
Film/Sheet (Gauge-mil)	2-50	2-5	1-3	5-125	1-15
Accuracy (Gauge Control)	3%	5%	10%	10%	7%
Start-Up Time (hrs.)	5-6	5-6	3-5	3-5	1
Wind-Up Speed (Ft/Min)					
(Avg)	240	180	45	45	60
(Top)	450	240	80	90	120
Limitations	High capital cost	Lower rate, versatility	Poor accuracy. Reduced versatility	Low rate	Inefficient, High energy cost-release paper cost
Advantages	Versatility High rate Accuracy	Accuracy Reduced control	Low investment Thin Gauge (3 mil and under)	Heavy gauge (5-125 mil)	

Source: Plastics World, July 1979

IRRADIATED POLYETHYLENE

The effect of irradiation on polymeric materials has been widely investigated. One of the most successful uses of radiation has been in the crosslinking of polyethylene film. The properties of ordinary polyethylene film made by the bubble process are compared to those of irradiated film made by the same process in Baird's patent* (Table 21 and Figure 40). Cross-linking increases tensile strength, orientation release stress, and heat-sealing range. The toughness of the film is similar to that of Saran™ film and its low-temperature flexibility is superior. Over 90% of all frozen turkeys are packaged in films of this type.

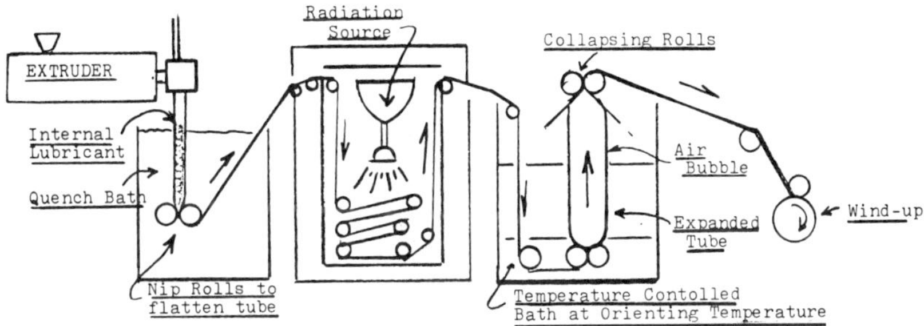

Figure 40. Process for biaxially orienting film of irradiated polyethylene.

Table 21. Effect of Irradiation on Properties of Oriented Polyethylene

	Irradiated	Not Irradiated
Specific Gravity (g/cc)	0.916	0.916
Tensile Strength, psi		
At 22°C	8,000-16,000	1,500-3,000
At 93°C	1,500-3,000	100-200
Elongation (%)	100-200	50-600
Heat-sealing range, (°C)	150-300	110-150
Shrinkage at 98°C (%)	20-80	0-60
Orientation release stress at 96°C (psi)	100-500	0-10

Courtesy of Cryovac Divn. of W. R. Grace & Co. U.S.P 3,022,543 W.G. Baird et al.

This film, made by Cryovac™ Division of W. R. Grace, is applied in shrink wrapping, particularly where its good low-temperature properties can be used to advantage. This process is essentially identical to the original Saran™ film bubble process, described in Figure 23 with the dif-

*See Appendix #13.1.1, No. 5, 8

ference that the extruded, quenched, thick-walled tube is irradiated before being oriented in bubble form. The irradiation step cross-links the polymer molecules to such an extent that the film can be stretched but no longer becomes fluid at its original melting point of 105-110°C. Thus, it retains a relatively high orientation release stress at 100°C, producing tight shrink wrapping in boiling water. This is why cross-linked PE is used for tough long-term applications like frozen fowl while unirradiated oriented LDPE can only be used for short-term, milder applications, e.g. shrink wrapping skids or boxes, or shrink sleeve wrapping of containers on corrugated trays.

Special copolymers of polyethylene, specifically those that enhance melt elasticity and broaden the orientation range, have been of interest to those wishing to reduce the cross-link density requirements. For example, as one increases the long chain branching in ethylene polymers (i.e. by copolymerizing with longer chain alpha olefins) one can broaden the orientation temperature range or produce cross-linked systems having equivalent shrink properties at lower cross-linked density. Copolymers of PE/EVAC make excellent shrink films with very little cross-linking. Early work on these systems led to the development of modern-day stretch film grade resins.

Other approaches toward improving shrink properties of oriented films have been to "build into the molecule psuedo-cross-links." DuPont has done this with their original Clysar™ resins which were blends of high and low density polyethylene. The high density polyethylene served as a reinforcing material and its crystalline nature acted as a "thermally liable" cross-link. Other approaches have been to use chemical cross-linking to achieve the degree of orientation with adequate shrink properties as an alternative to radiation cross-linking.

Radiation cross-linking has two major advantages: FDA approval and no residual contaminants.

HIGH DENSITY POLYETHYLENE (HDPE)

Markets & Supply – Between now and 1983 it appears that there will be an overcapacity of HDPE production. Figure 41 illustrates the relative supply and demand balance of HDPE. (Note: these figures *do not* take into account any corporate plans to switch HDPE production to LLDPE or to use present facilities to make HMW-HDPE resins). The continuing problem is the projected demand (from marketing professionals) does not

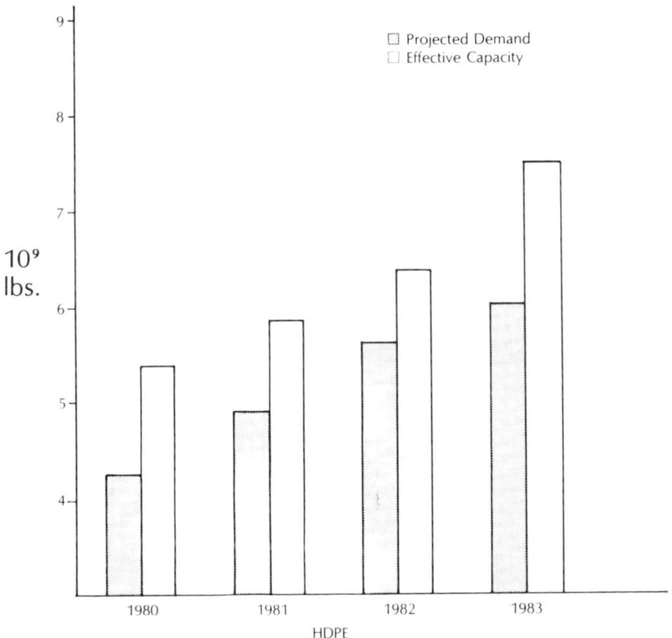

Figure 41. Projected supply-demand balance for HDPE through '83. (Dean-Witter Reynolds-Plastics World - Apr. '81). Note: Actual sales of HDPE in 1980 – 3.75 x 10^9 lbs. 1981 = 4.0 x 10^9 lbs. (Modern Plastics 10/81)

coincide with actual sales (Figure 41). However, based on published data (Figure 42) the industry could have a supply problem by the end of 1982, (but it's doubtful,) resin prices will probably increase in mid to late 1980's The one controlling factor may be the drastic decrease in exports which could leave more capacity for domestic markets.

Shrink Film Potential – HDPE has a melting point of 134°C, a glass transition temperature of −120°C, and is highly crystalline. The usual processing sequence of melt extrusion followed by quenching to give either an amorphous or slightly crystallized material for subsequent stretching is not practical with the high-density polyethylene resins because of the very high rate of crystallization, even at 0°C. Once crystallized it cannot be stretched very far without rupture. Thus, if this material is to be biaxially oriented from a melt, it must be done at the same time that it is cooling and crystallizing. Lubricants, crystallization in-

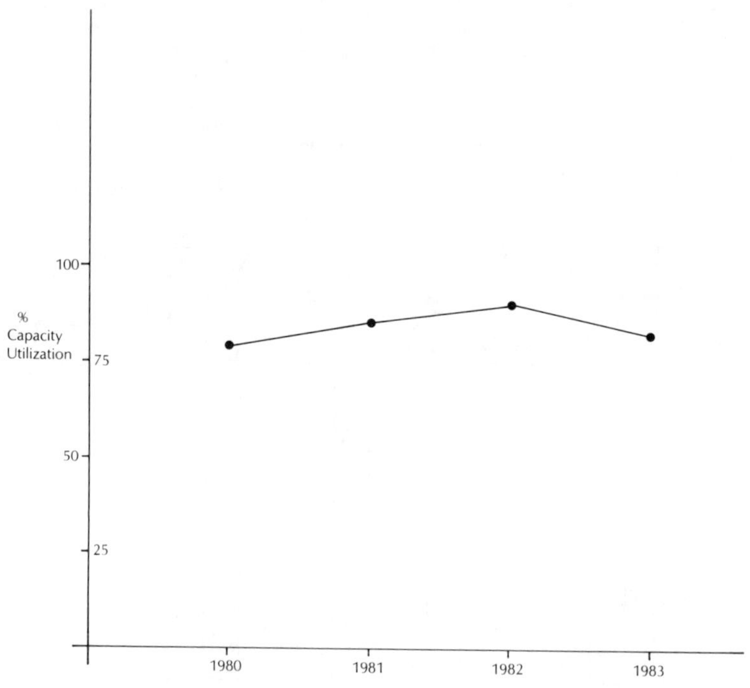

Figure 42. HDPE projected supply-demand. (Ref. 13.1.1, No. 3, 15,30)

hibitors, and correct time-temperature conditions to get acceptable rheology are required to orient materials of this type. (Previous work indicates that mixtures of high density PE and elastomeric polymers produce orientable mixtures).

One grade of uniaxially oriented HDPE shows *30% shrinkage* in the machine direction after 10 minutes at 265°F and *10% elongation* in the transverse direction. This shrink behavior indicates that the film was made by accelerating a sheet from a slot die and rapidly quenching. Since the transverse direction expanded on heating, it must be necked down during orientation. This same process, modified so the accelerated film wraps around a laterally expanding roll in a quench bath, has been claimed to be useful in partially biaxially orienting high-density (0.950-0.960) polyethylene. Only a limited degree of transverse stretching is claimed.

Work has also been done on a more complex tube-stretching apparatus, which grasps the edges of the hot extruded tube, draws them over a cooled mandrel which is mounted very close to the die, and then pulls the film away. Tensile strengths of up to 20,000 psi have been reported with polymers having a density of 0.960 at takeoff speeds of 12 ft/min.

Another modified bubble process which is said to be effective for the biaxial orientation of high-density polyethylene has an internal cooling mandrel the same diameter as the extruded tube. After cooling, the tube is reheated to the desired stretch temperature by the use of both internal and external radiant heaters. The tube is expanded, by air pressure after heating, and the amount of expansion is limited by contact with a cylindrical cooling jacket. Tensile strengths of 12,500 and 13,900 psi in the machine and transverse directions, respectively, are reported by this technique. (Shrink tension can be low in these cases).

However one film on the market is based on the partial solution of this problem. Addition of LDPE retards crystallization of HDPE, and DuPont's shrinkable polyolefin film is an excellent example of this. DuPont's chemists and engineers teamed up to make a shrinkable film of moderate shrink tension by a one-step bubble process. The film is approximately a 70/30 mixture of high density PE and low density PE. This specific formulation was developed to optimize the rate of crystallization, melt elasticity and rate of orientation to produce a clear, high strength polyolefin shrink film.

In summary, it appears that to date no commercially practical equipment has been developed for fabricating a true biaxially oriented high-density polyethylene film without modifiers. Those systems that have been successful incorporate resin modifiers or copolymers. (These systems do not have the properties of the irradiated or cross-linked polymer systems noted previously).

New Developments – The plastic industry has long envied the dominance of paper products, especially in the area of packaging. High density polyethylene has been a prime candidate to compete with Kraft paper in grocery sacks an bags. The amount of HDPE being sold in the form of grocery sacks is difficult to determine. This is due, presumably, to the proprietary nature of the information which includes marketing strategy, specific resins and bag design. We do know that approximately 100 thousand metric tons of HDPE went into bags and sacks in 1981. These figures, however, do not include the marketing of HMW-HDPE grocery sacks (Table 22). 1980 saw the first major jump in HDPE film in merchandising bags. The HMW-HDPE (or mixtures) grocery sack is the next target. Based on the attention this resin is getting, one can assume that the successful use of HDPE in grocery sacks or in the flexible film area will depend heavily on new resins and blown-film techniques taking place.

Table 22. HDPE Sales in Major Film Markets[1] (in 1000 metric ton units)

Markets	1980	1981
Merchandise Bags	30	65
T-Shirt Sacks	?	1
Trash Bags	?	4
Food Bags – Bag-in-Box Applications	25	22-23
In-store Bags & Wraps (Fish & Deli)	5	7

[1]Figures do not include 1/6 barrel, gussetted stand-up grocery sacks and others based on HMW-HDPE, competing with Kraft Paper.
Courtesy of Plastics World and Modern Plastics

This work is easily justified, when one considers that HDPE tonnage (*in film*) grew 50% the first three months of 1979 (compared to the same period in 1978) and is expected to continue at an increased *annual* rate of 9-10% per year. By 1983 tonnage will increase to approximately 100,000 to 300,000 tons for HDPE films. (In 1981 HDPE sales reached 100,000 metric tons excluding HMW-HDPE markets).

Where will the new generation of resins come from? And where will they be used?

Merchandise bags and grocery sacks are obvious uses for HMW-HDPE, *but watch the MHMW and HMW copolymers, these could be new shrink film resin candidates – they have the potential flow and crystallization rate properties.* Also, HDPE will probably replace paper in some applications before it replaces LDPE. Cost is the factor. (The bag-in-box application for dry foods is growing at an exceedingly fast rate. The question here is how much did the 1981 recession hurt market development?)

New Materials – At the present time, suppliers receiving the most attention as new generators of blown film resins are: Gulf, DuPont, American Hoechst and Union Carbide, according to Modern Plastics' August 1979 article and data coming from the National Plastic Exposition. In addition Soltex, Chemplex, Phillips Chemical, Dow Chemical, Arco, Allied, USI and Cities Services are introducing competing resins. (In 1981 Soltex announced new HMW film grade resins to compete with U C & C in grocery sack applications).

Other "blown film" HDPE resin types include *Medium molecular weight (MMW)* polyethylene homopolymers (i.e. d=0.96 min.melt indexes 0.8 to 1.0) which possess "paperlike" hand, stiffness, ease of processing and barrier properties.

Medium high molecular weight (MHMW) copolymer technology is imported at the present time from Europe and Japan. Two main reasons have been quoted for this phenomenon. (1) Lack of facilities and capacity and (2) fear of "down grading" and customer resistance to thin gauge "flimsy" films and bags. (Modern Plastics 8/79 - pg. 31). American Hoechst has been the only domestic supplier. (This will change in the 1980's with Gulf, DuPont and Union Carbide planning productions).

What supliers are committed to this market?

> *Soltex* will sell 2 to 4 new HMW and MHW-HDPE resins in 1981.
> *Chemplex* has issued reports that they will develop resins for blown film processing. *MHMW* copolymers, *MMW* homopolymers and *HMW* copolymer.
> *Phillips Chemical* has commercialized a merchandise bag resin (MHM-TR-140) - *MHMW* (but may delay heavy expansion plans).
> *Dow Chemical* is still a steady supplier and converter of cast HDPE film and is developing blown film candidates.
> *Arco, Allied, U.S.I.* and *Cities Service* will have *MHMW* grades in 1983 (but construction or modification of facilities is in question in 1980-1982).

New Equipment – At least three companies are developing new equipment to make extrusion-blown film.

A partial list of some of the activities described in various periodicals and new releases are:

> *Sterling Extruder* – Increased cooling, lowering frost line, new double flight screw design, low shear spiral die design and new tower.
> *Brampton Engineering* – Spiral flow die design, conversion systems (screw, die and cooling ring, collapsing assembly).
> *Sano Design and Machine* – Modified die, air ring (dual lip).

Internal bubble cooling is also a major activity with such equipment companies as Egan, Windmoeller and Hoelscher, and Brampton receiving some attention.

Summary & Future – In summary, these developments in both materials and processes definitely point to a major increase in HDPE technology. The prospects are bright that HDPE markets will continue to expand. At this point shrink and stretch applications are not attainable with homopolymers. The copolymers and polymer blends are fruitful areas.

The new ethylene-hexene copolymers from Phillips look like good shrink film candidates. Economics will dictate the future but paper replacements look promising. LDPE replacement is doubtful.

LOW DENSITY POLYETHYLENE & COPOLYMERS (LDPE)

Low density PE has been the dominant packaging film. High density polyethylene, on the other hand, has limited use in packaging (but is growing in the area of tissues, glassine and grease proof paper replacements). HDPE is not a significant factor in shrink or stretch wrap.

Low density PE film packaging applications grew by 13 to 15% (to approximately 1 million tons) in 1977.

Perhaps the most noteworthy polyethylene shrink film (outside of the radiation cross-linked film made by Cryovac™) has been a DuPont film that was developed in the early 1960's.

The DuPont filmis made from a resin blend that consists of 70% LDPE and 30% HDPE and is extruded and oriented on special bubble apparatus. The HDPE acts as a pseudo-crosslinking agent through the formation of crystalline regions. By carefully controlling the temperature of the tube, a high degree of stress is "frozen" into the film during orientation by the bubble process. Orientation is reported to be threefold in each direction at temperatures between 90 and 130°C.

Belgium patent 619,351 notes that 75% Alathon 1413 with 25% Alathon 7020 (with appropriate stabilizers and anti-block agents) produces resin with density of .925 and a M.P. 125°C.

This mixture is extruded through an annular die. The tube is quenched to room temperature with internal mandril cooling, then reheated to 115°C and blown up to five-fold its original dimensions in each direction. The final film is clear and shrinks between 15 to 25% at boiling water temperatures.

DuPont's CLYSAR™ family of resins consists of:

EH – a biaxially oriented PE
EH-F – an untreated grade of biaxially oriented PE.
EH-FT – an oriented PE film treated for ink receptivity and printing.

EH-T – biaxially oriented PE treated on one side. (This PE film is used in a variety of applications, specifically, overwrapping, bundling, window packaging, and tray wraps.)

EH-C – a biaxially oriented shrinkable copolymer film

ECL – is a cross-linked ethylene and this film is presumably similar to the Cryovac™ radiation cross-linked polyethylene. These films are noted for their strong heat seals and good shrink characteristics, although a comparision of the published data indicates that Cryovac™ film has more shrink and shrink energy. (These properties are discussed in a later section.)

Union Carbide and Exxon also market a series of polyethylene shrink films that are engineered for specific needs. Some U C & C film data is shown in Table 23.

Table 23. Polyethylene Shrink and Stretch Film Properties*

TYPE	Modified "high cling"	High slip anti-block	Low slip Cling
APPLICATION	Pallet Stretch	Pallet Shrink	Pallet Stretch
Density (gm/c.c.)	0.921	0.922-0.923	0.928
Modulus (1% Elong.)	30,000	28,000	21,000
Tensile st. (p.s.i.) MD/TD	2600/2000	2800/3000	3800/2800
Elongation (%) MD/TD	275/400	300/500	300/600
Haze (%)	–	58	75

Source: Plastics World and UC&C Data.

Some of the more popular commercial films are:

High clarity, high slip film for bundling and overwrap of industrial and consumer goods.

Low slip, sleeve wrap film that is directed toward high speed bundling machines (in areas of increased stackability).

Medium low slip, sleeve wrap film that is used for ice cream, beverage, can, bottle, or case shrink wrapping.

Medium clarity and haze with high strength performance for pallet wrap (bags and films).

Union Carbide's new low pressure, gas phase polyethylene process (licensed to Exxon and others) is expected to improve the economics of LDPE. However, early work indicates that we cannot expect significant upsets to develop in the shrink film market due to this resin appearing on the market at this time. On the other hand, stretch film is a fruitful area of investigation.

The packaging market for LDPE films consists mainly in bread and produce bags (non-shrink), meat and poultry (cross-linked PE shrink film and EVA stretch), industrial liners and general merchandising bags (non-shrink) and case wraps and pallet wraps (stretch and shrink) (Table 24).

Table 24. Polyethylene

ADVANTAGES	Low Cost Good Tear Low Temperature Properties Special Shrink Grades	
APPLICATION	Bags Product Packaging	Overwrap Agriculture Drum Liners
	General Purpose Barrier Fresh Meat Bakery Products Frozen Food	Shrink Wrap Cured Meat Milk

Using shrink or stretch film to unitize containers on a corrugated tray is now economically competitive with a corrugated box.

Pallet wrap (stretch and shrink) presently is a small part of the LDPE film market but the area is growing rapidly. Pallet shrink wrap films are usually made from LDPE's with density ranging from 0.918 to 0.930 gm/cc with a melt index on the order of 0.1 to 0.5. Tray wrap, on the other hand, is lower molecular weight (melt index greater than 0.7 to approximately 2.0) and a density of 0.920 to 0.925. Pallet stretch film grade resins feature toughness, extrudability, and cling (as desired). The multiple wrap material has melt indexes 1 to 6 and densities of 0.924 g/cc.

Although low-density polyethylene films made by the bubble process are generally not regarded as oriented, they shrink (up to 75%) when heated without restraint. Because they do not start to shrink until a temperature close to the melting point is reached, they exhibit very low or zero orientation release stress. At these temperatures, the polymer is nearly fluid since most of the crystallites have melted. Some uses have developed for these shrink properties, and because low-density polyethylene is the least expensive polymer film, formal attention is being paid to modified bubble processes which control the shrinkage characteristics of this film.

Structure and Processability – As noted in other sections, the process of orientation imparts potential shrinkage to a polymer film by aligning the molecules and freezing the molecules in the aligned or "stressed" state. The process consists of stretching the plastic melt and cooling the film. Orientation of LDPE takes place between 195-204°F. When the polymer is quenched the molecules are frozen in the stretched position. During the shrink process, when the film is re-heated above the softening temperature the molecules are free to return to their original random configuration.

Melt viscosity, melt elasticity, density and rate of crystallization are characteristic properties that determine or influence the orientation process and the final shrink properties. These can be translated into polymer properties:

> A high *molecular weight* resin is preferred for increased melt strength and recoverable shrink.
>
> The *molecular weight distribution* should be narrow. This insures a more uniform melting and shrink throughout the polymer mass. As a result the film requires less heat and less time to reach maximum shrink.

The rate of stretching (or orientation) determines the amount of energy or stresses imparted to the film. As orientation rate increases, the degree of stress frozen into the film increases. The recoverable shrink also increases. It must also be noted that orientation should be performed at the highest quenching or cooling rate so that crystalline growth is reduced and high degrees of shrinkage are maintained.

Therefore, in LDPE shrink films for packaging the important resin parameters are:

MOLECULAR WEIGHT
MOLECULAR WEIGHT DISTRIBUTION
LOW DENSITY (LOW CRYSTALLINITY)
LOW RATE OF CRYSTALLIZATION

The key parameters in the process are:

STRETCHING RATE & TEMPERATURE
DEGREE OF STRETCH
COOLING RATE (QUENCHING RATE)

These key parameters are converted into *actual process variables* according to experts in the field.* They have described the variables that are important in making LDPE shrink film in commercial quantities.

For example, the *degree of orientation* in blown film orientation is controlled by temperature, blow-up ratio, draw-ratio, and frost line height.

Blow-up ratio is determined by die opening, take-off speed and output. Draw ratio is the ratio of die opening/film thickness.

All of the orientation in the blown film process takes place below the frost line. This is the distance above the die in which the polymer is cooled below its melting point and thereby freezing in the orientation stresses. Reducing the height of the frost line locks in the orientation stresses at a faster rate. As the frost line height is increased, the chains have more time to relax and relieve the internal stresses. This results in less shrink, or shrink tension.

Generally, as blow-up ratio increases TD orientation increases and film thickness decreases. As take-off speed increases MD orientation increases.

In addition to blow-up ratio there are other process parameters that must be stressed:

Temperature
Processing temperature must be kept as low as possible to take advantage of the high melt viscosity and the increased level of stresses that can be retained.

Cooling Rate
This must be quick to retain the orientation stresses. (Low frost line and high extrusion rates result in quenched films with high degree of orientation.)

Bubble Shape
A mushroom shaped bubble in place of the usual inverted bell shape induces MD stretch prior to TD stretch and produces a more balanced film.

Package Integrity – Perhaps one of the most important functions of a shrink film is its ability to hold a shape with sufficient retractive forces, (shrink tension) to hold and protect the package and its contents. This means that retention of shrink energy, or creep resistance, is very impor-

*Appendix 13.1.1 #43 13.1.2.

tant. Very high molecular weight film resins and cross-linked films are creep resistant. (One excellent example is the turkey wrap and fresh meat packages available today.)

Several years ago Northern Petrochemical initiated a study to determine the elastic properties of different types of LDPE to identify those resins that were best suited for stretch film applications. (No work has been published to illustrate the shrink film requirements needed for a given package stability requirement.) A good balance of properties of LDPE are illustrated in Figure 43 and Figure 44.

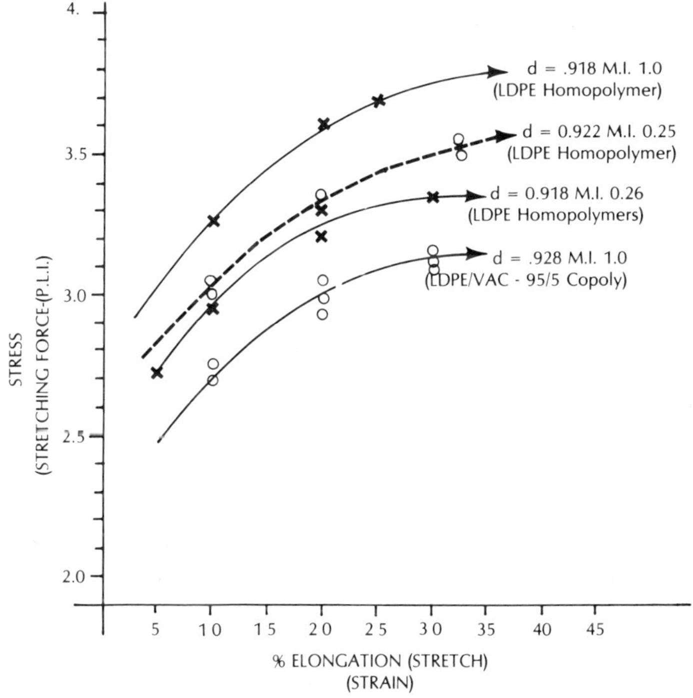

Figure 43. Stress-strain curves of 3 mil LDPE films, of different densities. Source: Packaging & Development – 11/12/74.

There is a serious lack of professional information in the literature concerning the evaluation of actual performance criteria of stretch and shrink film. For example, there are no references that describe the meaning of creep and its effect on the actual package integrity. Literature is also lacking that describes a standard (can be arbitrary but constant) shrink or stretch wrapped pallet and the performance of various films on that pallet, over various time frames. This is particularly important when discussing loss of tension at different stretch levels. Information on the

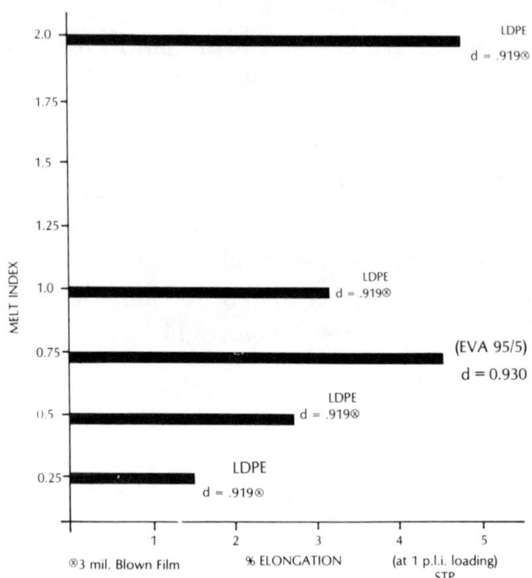

Figure 44. Creep of LDPE film. Source; Package Development – 11/12/74.

decrease of retained stress with film structure, process variables, and film thickness is also lacking.

For example how does one transpose the data in Tables 25 and 26? Does 95% recovery in 24 hrs. mean that we will lose 5% of our "holding power" in 24 hours? If this is so, what happens to a pallet on a truck going across country? What are the time limitations on thermoplastic films used in stretch or shrink applications?

Clarity of information is important from the standpoint of engineering a film to do the prescribed job without using film that is too thick or has too many winds. (In other words, over engineered or optimized protection.)

LINEAR LOW DENSITY POLYETHYLENE (LLDPE)

In 1977 Union Carbide introduced a new linear low density polyethylene that was produced in a gas phase polymerization reaction. It was a copolymer of ethylene and butene. Shortly thereafter, Dow Chemical Co. announced a solution process for making a similar LLDPE. The Dow process is reported to be an adaptation of their HDPE process and produces an ethylene/octene copolymer. DuPont of Canada is also producing LLDPE by solution process. This is reported to be an ethylene/hexene copolymer.

In 1981 the potential for producing 450,000 metric tons of LLDPE was

Table 25. Creep in Machine Direction (24 hour test. 3 mil. film; 2.2 BUR*)

Melt Index	Density	% Creep
0.25	0.919	2.0
1.00	0.919	3.0
0.25	0.919	2.0
0.25	0.922	1.75
0.50	0.919	2.5
1.00	0.919	3.0
2.0	0.919	4.5
1.00	0.919	3.0
1.00	5% Vac	5.0

*BUR – Blow-up Ratio

Source: "Package Development Nov.-Dec. 1974 "A Study in Resin Types by D. Rex Young.

Table 26. Elastic Recovery (%) (As a Function of Molecular Weight) (3 mil PE film - 0.919 - BUR - as above)

Melt Index Elongation	Elongation 10%	20%	30%
0.25	97.6	95.7	94.3
0.5	97.6	95.6	94.2
0.75	97.6	95.6	94.2
1.00	97.6	95.5	94.1
2.00	97.6	95.5	94.0

Source: "Package Development Nov.-Dec. 1974 "A Study in Resin Types by D. Rex Young.

in place in the U.S. Although potential expansion figures may be optimistic, publications (Appendix 13.1.1) indicate that DuPont's Canadian commitment will total 450,000 metric tons/yr. Union Carbide- 620,000 metric tons/yr.- Dow Chemical - 515,000 metric tons/yr. (includes 200,000 for Benelux). Union Carbide's licensees include Mobil (140,000 metric tons/yr.) and Exxon (275,000 metric tons/yr.).

In addition, announcements of major developments in making LLDPE have also been published for the following:

COMPANY	PROCESS	COPOLYMER
U.S.I.	Slurry Process	Ethylene/Butene
El Paso Polyolefin	"Liquid Pool"	Ethylene/Butene
Phillips Chemical	Modified Particle Form Process	Ethylene/Butene
ARCO	Medium to High Pressure	Ethylene/Butene
Cities Service	Medium to High Pressure	Ethylene/Butene

Structure and Properties – LLDPE is a high melting (approximately 20°C above conventional LDPE) linear polyethylene with a density equivalent to LDPE, and 10% more crystallinity due to its more "ordered"structure. LLDPE has a narrower molecular weight distribution and is lightly branched (due to the olefin comonomer, the catalyst structure, and the mild polymerization conditions). Chain branching is essentially random. Figure 45 graphically illustrates the two LDPE's.

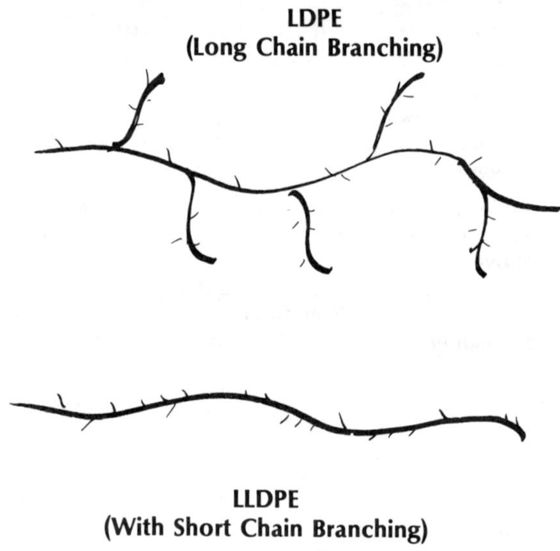

Figure 45. Dow and Union Carbide Data supplied by companies.

The properties of LLDPE are what one would expect from a linear, more crystalline polymer of narrower molecular weight distribution. For example, improvements are noted in impact strength, toughness, environmental stress-crack resistance, modulus (higher) and resistance to tear propagation. Optical properties suffer and processing is more difficult. Processing LLDPE involves a higher operating temperature (higher melting point), increased melt viscosity, and less shear sensitivity.

The film properties of LLDPE are less sensitive to relatively large changes in blow-up ratio. In addition, LLDPE resins are not generally recommended for shrink film applications. Orientation is more difficult to obtain due to absence of long chain branching. This means that the

LLDPE's have a much faster relaxation time than LDPE, due to the presence of short-chain branching. They are not candidates for shrink film application (i.e., rate of relaxation is too rapid). These same properties, (especially in the cast film) make LLDPE resins ideal stretch-wrap candidates. The data in Tables 27 and 28 (depicting two major sources of LLDPE) show the major properties closely allied to stretch, i.e., high elongation, and tensile strength, and resistance to puncture and tear propagation. (Compared to LDPE and HDPE, LLDPE appears to be the best candidate (Table 29).

Table 27. Blown Film Mechanical Property Data (Source: Union Carbide)

		Union Carbide LLDPE		High Pressure LDPE			
Melt Index (Gm/10 min.)		1.0	2.0	2.0	2.0	0.2	0.1
Density (Gm/Cm³)		.920	.920	.923	.918	.923	.923
Dart Drop (Gms)		145	90	90	112	185	200
Puncture Energy (in-lbs./mil)		15.3	14.6	7	6	4.5	4.0
Elmendorf Tear (gm/mil)	MD	120	99	260	160	90	50
	TD	340	245	160	110	100	100
Tensile Strength (psi)	MD	5880	4970	2700	2900	2800	3050
	TD	4660	3770	2400	2700	3000	3000
Tensile Elongation (%)	MD	620	690	350	300	300	370
	TD	760	738	500	500	500	540
Tensile Impact Strength	MD	1241	858	440	440	480	490
(ft.-lbs./in.³)	TD	674	433	650	650	1030	1100
Secant Modulus (psi)	MD	34000	30600	24600	21200	23000	25200
	TD	37500	37000	27000	25700	26000	28000

Conditions: 1.5 mil gauge; 2:1 BUR.
Ref: Film Extrusion of Low Pressure LDPE by W. A. Fraser, L. S. Scorola & M. Concha Polyolefin Div. -UC Corp.

Packaging film applications require specific properties. These are Dart Impact, Elmendorf tear and ultimate tensile strength. Films fabricated from LLDPE have significantly higher impact, tear, and tensile strength than do films of the same molecular weight (as measured by melt index) and density. Because of the improved physical properties, one can either fabricate films of equivalent thickness with superior properties or thinner films with equivalent properties. Data indicate that one can realize as much as 20-25% savings in resin at equivalent impact and tensile properties.

Table No. 27 compares the mechanical properties of blown film made from conventional *high-pressure* LDPE and *low pressure* LDPE made by Union Carbides' gas phase polymerization process – Unipol™.

Table 28.
Fabrication Conditions for Tubular Film Extrusion:
Melt Temperature = 450-500°F
Blow-up Ratio = 2:1
Optimum Gauge Range = 0.5-3.0 mil

Physical Properties[1 2]		**ASTM Method**	**Value**	**Dowlex™ LLDPE**		
Melt Index, gm/10 min.		D-1238	0.8	1.0	1.0	1.0
Density, gm/cc		D-792	0.930	0.935	0.926	0.920
Vicat Softening Point, °C		D-1525	112	118	106	100
Tensile Yield[3], psi		D-638	2400	2800	2200	1800
Ultimate Tensile[3], psi		D-638	4000	2500	3300	3500
Ultimate Elongation[3], %		D-638	900	800	800	1100
Tensile Modulus[3], 2% Secant, psi		D-638	52,000	61,000	56,000	40,000
Film Properties[1] @ 1.5 mil						
Dart Impact, gm		D-1709	190	110	210	250
Elmendorf Tear Strength, gm	MD	D-1922	230	100	240	450
	TD	D-1922	600	410	430	680
Tensile Yield[3], psi	MD	D-882	2200	2320	1900	1500
	TD	D-882	2350	2640	2000	1600
Ultimate Tensile[3], psi	MD	D-882	5400	5700	5300	5000
	TD	D-882	5500	4400	5100	4500
Ultimate Elongation[3], %	MD	D-882	660	740	700	750
	TD	D-882	740	770	750	750
Gloss, 45°		D-2457	45	20	37	37
Haze, %		D-1003	13	30	16	16

[1]Typical properties; not to be construed as specification limits.
[2]Compression molded samples.
[3]Crosshead speed – 20 in/min.
Source: Dow Chemical Data

Table 29. Comparision of Polyethylene Film Properties (Sources: Exxon/Plastic World)

Resin	LLDPE	LDPE	HDPE
Haze	Medium (blown) Very low (cast)	Low	High
Tensile strength	Medium	Low	High
Elongation	Very high	Medium	High
Tear resistance	Low-High	Medium	Low
Impact strength			
Room Temp	High	High	High
Low Temp	High	High	High
Puncture resistance	Very high	Low	Medium

Source: Plastics World: Exxon Data

Table No. 28 on the other hand compares the mechanical properties of compression molded and blown film samples made from Dow Chemicals solution polymerized low-pressure LDPE. Dowlex™. Dow also has published reports that a new class of Dow LLDPE resins (presumably ethylene - butene copolymer) can be made in their prsent high pressure LDPE facilities using a proprietory catalyst and lower pressures - (All Dow resins will be sold in pellet form.)

New LLDPE Technology – In October 1981, Union Carbide delivered the first barge-mounted modular LLDPE plant with a design capacity of 240 million pounds per year. The plant was fabricated in Japan and delivered to Balia Blanca, Argentina in 21 months. Union Carbide has called the concept the W (Waterborne) Plant Option. (Some years ago a barg mounted pulp mill was floated up the Amazon in Brazil).

In the fall of 1981 other LLDPE processes were being announced.

El Paso Polyolefins and Arco Polymer announced new processes. El Paso, using a new high yield catalyst, will be able to make LLDPE in its Odessa polypropylene plant without losing polypropylene production. The new technology from Montedison will improve the yield of the "liquid pool" process. (This ethylene - butene process is being licensed). Arco has also announced the pilot plant production of these resins.

Phillips Chemical Company is building a new LLDPE facility to produce 8 million lbs per year. This technology, originally investigated in the late 1950's and early 1960's, is based on modification of their "high yield" particle form catalyst. Market predictions are that Phillips will license the technology to its present HDPE licensees to make either HDPE or LLDPE in their present facilities with relatively minor modifications.

Impact of Technology – The growth of LLDPE and its acceptance in the market place has been astounding – from the Unipol™ process announcement by U C & C in 1977 to an in-place production capacity of 450,000 metric tons per year by 1981, to a predicted 2 to 3 billion pounds (1.36 million metric tons) in 1983 (Figure 46). It also has raised the question of over capacity (Table 30).

As of 1981 LLDPE production accounts for 5% of the total LDPE production. By 1985 it's expected to be 25-30% of the LDPE production and markets.

This expansion of production capacity is based on some economical "facts of life."

> The *improved processing* (melt drawdown) allows thinner films to be made (at lower raw material cost) with equivalent properties to LDPE. This allows thinner films (alone or in coextrusion) to be used in composites. Several excellent examples of this down-gauging were presented in the Feb. 1981 issue of Plastics World (Table 31). Down-gauging can be accomplished

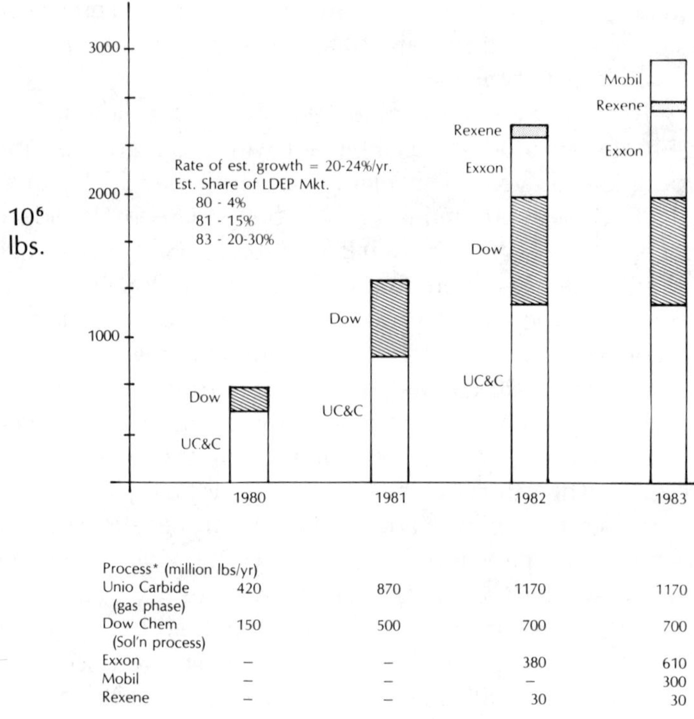

Figure 46. LLDPE estimated maximum operating rates. (Source: Plastics World – Dec. 1981 - pg. 72).

Table 30. LLDPE Nameplate Production Capacity, Consumption, and Calculated Overcapacity

	Capacity (lbs/yr)	Consumption (lbs/yr)	% (lbs/yr)	Excess Cap. (lbs/yr)
80	570×10^6	250×10^6	43.8	320×10^6
81	1370×10^6	700×10^6	51.2	670×10^6
82	2280×10^6	1500×10^6 (est)	65.8	780×10^6
83	2805×10^6	2125×10^6 (est)	78.4	680×10^6

Source: Plastics World Dec. 1981 pg. 72.

Table 31. Gage Reduction Potential For LLDPE Bag Applications (1,2)

LDPE (Products)	LLDPE (Products)	Reduction (%)	Applications
2 Mil	1.3 Mil.	35%	Single Ply Trash Bags
1.5 Mil	1.3 Mil.	13%	Conventional Trash Bags
2.0 Mil (EVA-LDPE)	1.5 Mil.	25%	Ice Bags
1.75 Mil	1.25 Mil	29%	Tape Drawstring Bags
0.9 Mil	0.75 Mil	22%	Product Bags

1)Clarity is not a required property in these applications. If clarity is required, than 80/20 or 50/50 blends of LDPE/LLDPE would reduce the saving accordingly.
2)Source: Exxon Chemical – Plastics World, Feb. 1981.

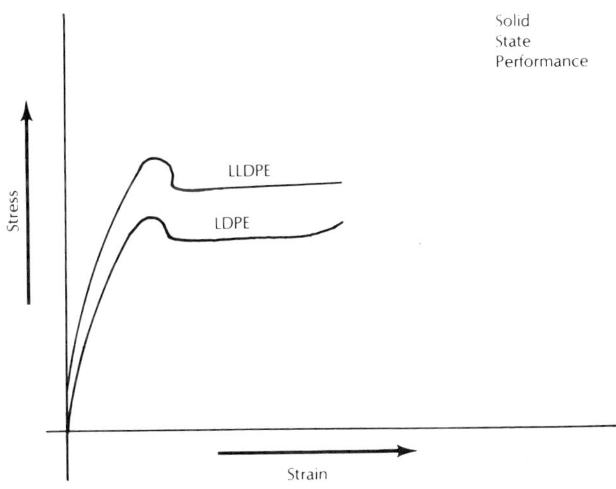

Figure 47. Solid state performance.

because LLDPE can be extended with relative ease (it is less sensitive than LDPE to large deformation) – so much so that thin films (.00025″) can be prepared without loss or rupture of the bubble (in blown film extrusions). Modified air rings are usually used to ensure the stability of the bubble – for example, air collars or baffles are used to direct the cooling air along the surfaces in the MD. The extension behavior of LLDPE relative to LDPE was described in a paper given at the 1980 TAPPI "Paper Synthetics" Conference by Union Carbide personnel. (Appendix 13.1.1).

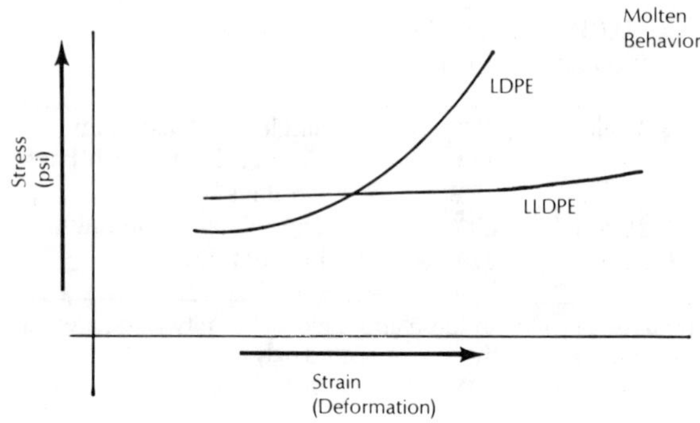

Figure 48. Melt extension: Source: Exxon and UC&C data.

A comparision of the LDPE and LLDPE stress-strain properties presents the key (Figures 47 and 48). During drawdown or melt extension, the conventional LDPE increases its apparent melt viscosity. This extensional viscosity is more constant for LLDPE or, as noted previously, LLDPE is less shear sensitive.

The potential for cost savings is another driving force, especially in the cost of raw materials. Conventional LDPE is produced in expensive, high pressure (30 to 50,000 psi) reactors, at temperatures up to 300°C and using.a peroxide catalyst. LLDPE can be made in present HDPE, and PP reactors at conventional (medium to low) temperature and lower pressure.(UC&C – gas phase process; Dow – solution process; Phillips – particle form process and El Paso – liquid pool-Ziegler/Natta process). These processes take about 1/3 the energy and 1/2 the capital of a new plant (or as Dow noted – 10% of the capital to convert a present facility). All these facts make LLDPE production attractive and conventional LDPE facilities financially unattractive.

Applications – There is no doubt that LLDPE has great potential because of low cost and improved properties. However, there are some deficiencies that present a challenge. LLDPE is a good blending resin and can be blended with LDPE to improve its performance, but the clarity of LLDPE is poor and the blown films need additives to improve optical properties. Processing of LLDPE is more difficult because it is more crystalline and linear. Modified dies and screws are required to obtain acceptable lbs./hp/hr. which are 20-30% less than standard LDPE. Table 32 lists some of the advantages and disadvantages of LLDPE.

Table 32. LLDPE (Film)

ADVANTAGES

Excellent strength properties:
- Tensile
- Elongation
- Tear
- Toughness

DISADVANTAGES

Optical Properties:
- Blown film – low
- Cast film – acceptable

APPLICATIONS

Coextrusion
Bags (ice, trash)
Frozen Food Packaging
Bakery Film

Cast Film
Diaper Liners
Stretch Film

Blown Film
Heavy Duty Bags
Grocery Sacks (with LDPE or HMW-HDPE)
Produce Bags

In spite of these shortcomings, application research indicates that about 70% of the LLDPE resin will initially go with film applications (alone, as a blend, and as a co-extruded layer in a composite film) and will definitely influence the packaging film market. Some of these applications are listed in Table 33. (The reader will note that most of these applications do not require optical clarity).

One approach being commercialized is to improve the optical properties by blending or coextrusion.

LDPE/LLDPE/LDPE coextruded film will give better physical properties and clarity than 100% LDPE.

Coextruded LDPE/LLDPE is being used in heavy duty sacks. (Impact strength of LLDPE is retained and elongation under load is decreased).

Coextruded LDPE/LLDPE is also used in medical packaging due to the excellent integrity of the heat seal of LLDPE.

Conversion Costs and Efficiency – Extra conversion cost is required if one is to use granular LLDPE produced by the Union Carbide Unipol™ process in standard LDPE material handling lines (designed for pellets). Modern Plastics, November 1981 issue notes that estimated costs could range between $5,000 and $20,000 per line.

Table 33. LLDPE Film Markets & Applications

Market	Product Mix	Penetration (date)
Trash Bags	Coextruded Film LDPE/ LLDPE/LDPE LLDPE/LLDPE Blends LDPE	LLDPE/LDPE (70/30 by 1984)
Millinery Bags	LDPE HDPE LLDPE Blends	LDPE (>70% in 1984) LDPE (~20% in 1984) HDPE (10% in 1984)
Grocery Bags	Kraft Paper HMW-HDPE LDPE LLDPE Blends with LDPE & HDPE	Kraft Paper (93% to 96% in 1985) HDPE (HMW) (3% in 1985-1986) LDPE (1 to 2% in 1985) LLDPE (1 to 2% in 1985)
Produce Bag	LDPE LDPE/EA LLDPE Blends	LDPE (25% by 1984-1985) LLDPE (70-75% by 1984-1985)
Frozen Food Bags	Coextruded EVA/LDPE LLDPE Blends & Coex- truded Films LLDPE	LLDPE – 50% in 1984 LDPE – 50% in 1984
Bakery Film	LDPE LLDPE	LLDPE (40-50% by 1985) Based on cost, clarity and ability to run on bagging equipment
Heavy Wall Bags	LDPE HDPE (HMW) LLDPE Coextruded LLDPE/LDPE	(Industrial application) LDPE/LLDPE 50-65% in 1985 HMW-HDPE – ?
Multiwall Liners	LDPE LLDPE HDPE	LLDPE – 50-60% in 1986
Stretch Wrap	LDPE LDPE/EVA (Copol.) LDPE/LLDPE (Coext.) LDPE – LLDPE (Blends)	LLDPE – 50% in 1986
Corrugated Tray Wrap	LDPE/EVA LDPE/EVA LLDPE BLends	50/50 LDPE & LLDPE

Source: Plastics World Dec. 1981, June 1982, pg. 69-72; and Modern Plastics Dec. 1981, Jan. 1982.

The most effective way to incorporate additives is by compounding in the melt. (In the future, pelletized LLDPE's will be used alongside pelletized LDPE. Many producers of LLDPE pellets are now on-stream as of January 1982).

The key is in the cost.

Pelletized Dow resins sell at 3 to 4¢ per lb. premium over granular. The Unipol™ process is said to produce LLDPE granular at 6-7¢ per lb. less than solution resin. Considering the cost of blending and pelletization it could be a draw.

A summary of processing "pluses and minuses" is shown in Table 34.

Table 34. Processing Pluses and Minuses[1] (LLDPE vs. LDPE)

Advantages	Disadvantages
Less extension thickening	Higher melt flow temperature
High drawdown ratios attainable (100:1 vs. 40:1)	High melt viscosity
Wide die gaps	Lower extruder through put (HP/LB/HR)
	Less shear sensitive
	Resin plastication requires more energy
	Narrow die gaps cause melt fracture

[1]To assist converters – producers recommend shorter screws to improve output (L/D 18/1 vs. 24/1), die gaps of (50 mil. minimum), proper air cooling control and new screw designs.

A case study described in the April 1981 issue of Plastics World and using data supplied by N. Wheeler of Davis-Standard points out the problems of converting present LDPE film equipment to LLDPE.

As a general rule, the following points should be considered:

> If one wished to delay capital expenditures to convert LDPE facilities to LLDPE, one should consider a LDPE/LLDPE blend. Determine the appropriate composition and properties and compare the cost.
>
> One can increase the flexibility of blown film extrusion by having a set of replaceable inner die lips for the blown film die that can run LDPE and LLDPE.

Some specific recommendations described in the literature are:

LLDPE can be processed on a LDPE screw but at a loss of 20-50% in output (LBS/HP/HR).
(Recutting the screw increases efficiency by 25%).
Barrier screws can process LLDPE at 75-80% of LDPE rate.
Torpedos reduce HP/LB/HR, but also reduce output.

Die design is important in the extrusion of LLDPE to reduce melt fracture and improve output. During the extrusion of LLDPE, one must avoid higher pressure at the die that reduces output and encourage melt fracture.

Normally, die gaps of 20 to 35 mil are required for blown film extrusion of LDPE. *Granular* LLDPE requires 60 to 125 mil with faster take-off speeds. *Pelletized* LLDPE requires 50 to 70 mil.

LOW DENSITY POLYETHYLENE/VINYL COPOLYMERS

EVA Stretch Films – In the late 1960's a major polymer effort began that was directed toward the processing and the identification of commercial markets for ethylene copolymer films. Some of these specific efforts involved polymers containing vinyl acetate, acrylic acid, acrylic esters, anhydrides, and alkyl substituted monomers such as methacrylic acids and esters. Each of these systems has found its own niche in the packaging field. Perhaps the most significant system showing the most potential (in the shrink and stretch film area) is the vinyl acetate copolymer system, when applied to the packaging of meat and produce and the shipping of unitized packages.

The EVA copolymers are converted into film by standard high speed extrusion/blown film processes. High speeds are obtained through various methods of quenching the tube as it leaves the extruder and adjusting the temperature just prior to film formation or orientation.

The EVA films (ethylene vinyl acetate) have lower sealing temperatures, higher impact strength, and are tougher than low density polyethylene homopolymer films. As such they have potential in the areas where they impact on the applications currently enjoyed by the LDPE and PVC stretch and shrink applications. As one would expect, impact strength increases with vinyl acetate content, and molecular weight. The polymers become less crystalline, are more elastic, and possess the tendency to block at the higher VA levels. This means that one must seek a proper balance between the optical properties and "cling," since (as has been noted in many magazines like Modern Plastics) one must add anti-block agents or slip additives that tend to reduce sparkle, clarity and increase haze. In addition, as crystallinity decreases and VA increases, permeability to gases, moisture, oils and fats increases.

Ethylene vinylacetate copolymers have an excellent future. The ability to be stretched over a package, retain its tension over the use time of the item (package, pallet, or corrugated tray containing can, bottle or tissue) with minimum relaxation and FDA approval can make this family of films a potential marketing adversary. It is just a matter of time, process improvement, and marketing expertise before EVA polymers start pushing PVC stretch out of the food packaging area.

The success of ethylene vinyl acetate copolymers is due to several factors. For example:

The cost of energy is forcing industries to re-evaluate their heat intensive operations. While shrink operations are not as heat intensive as in former years they do require more heat than a stretching operation.

The cost of energy is also forcing industries such as producers of beverages, canned goods, metal and glass containers to review their shipping costs from different angles. Ethylene homopolymers and copolymers have shown excellent promise in protecting packaging units in transit, specifically when they are "spiral wound." (The key will be to find the optimum film thickness). In addition to cost, new polymers containing higher vinyl acetate (greater than 8%) in thin gauges (approximately .5 to .75 mil) look like prime candidates for displacing plasticized PVC in food packages, as we noted before.

DuPont's new EVA copolymers with 12% vinyl acetate look like good contenders. Tables 35 and 36 compare the new "high" vinyl acetate films with LDPE and PVC.

Ethylene/Methacrylic Acid Copolymers – A second family of ethylene copolymers has also shown exceptional promise, this is the family of Ionomer™ films used in coating and laminates. The Ionomers™ are compounded with various cations to produce materials that are essentially ionically cross-linked in the solid state but exhibit normal thermoplastic behavior in the molten state. This is due to the ionic cross-links that are thermally labile. Therefore, these materials can be processed on high speed thermoplastic processing equipment. Due to the ionic character of the polymer, these coatings and films show excellent adhesion to such substrates as aluminum, paper, and wood. They exhibit excellent tack strength (which is required for paper coatings and as an adhesive in structured films [laminates]). They also possess toughness (due to the

™DuPont Co., Inc.

Table 35. Wrap Force Retention* (Typical Values at Maximum Practical Stretch)

	PVC (0.8 Mil)	LDPE (1.0 Mil)	EVA 3135* (1.0 Mil)	EVA 3135* (1.0 Mil)
TENSION SETTING	23	38	33	19
WRAP FORCE (lbs)				
INITIAL	4½	7-1/8	8½	6-3/8
AFTER 16 hrs.	2½	6-1/8	7¼	5½
WRAP FORCE RETENTION (%)	56	86	85	86
STRETCH (%)	25-30	25-30	40-45	25-30
WEIGHT OF FILM REMOVED FROM PALLET LOAD (lbs)	1.09	1.04	0.90	1.10

*DuPont Values (3 wraps top and bottom)

	PVC	LDPE	EVA 3135
CLING (film to film)**			
90 degrees	110	50	150
135 degrees	130	60	180

**Force in grams to peel film from adjacent layer of film.
Source: DuPont Data – EVA3135X; Package Engineering July 1979, pg. 31.

Table 36. Effect of Tension on Yield and Performance of Stretch Films (PE/VAC 88/12 Copolymer)

Tension Setting: (PSI)	19	33	38
Wrap Force (lbs)			
Initial	6-3/8	8½	9¾
After 16 hrs.	5½	7¼	8¼
Wrap force Retention (%)	86	85	85
Stretch (%)	25-30	40-45	45-47
Wt of film removed from pallet (lbs)	1.10	0.90	0.70
Film cost (¢/lb)	85	85	85
Cling force (gms)			
Film to film			
90°	150	150	150
135°	180	180	180

Source: DuPont Data – EVA3135X; Package Engineering July 1979, p. 31.

ionic cross-links), low temperature heat seals, and oil, solvent and grease resistance.

At the present time DuPont owns the technology and name. There are approximately 24 grades of Ionomer™ resins for conversion into composite structures. Ionomers™ are used as heat seal layers, and as a coextruded film layer with nylon or polyethylene for controlled atmosphere packaging of sub-primal cuts of fresh meat in central distribution points. The resins are also housed as extruded coatings on aluminum foil to protect foil from pin-holing due to flexure, and supply an excellent heat seal layer.

Ethylene/Methacrylate – Strictly speaking EMA is not an accepted shrink or stretch film, but it is becoming an accepted laminated resin because of its flex strength and excellent heat sealing characteristics. The copolymer exhibits excellent adhesion to polypropylene, polyethylene and cellophane, in addition to paper and aluminum foil.

The resins are stable (to 600°F) so that they can be processed by extrusion, i.e. chill cast, coated, extrusion laminated, injection molded, and profile extruded. Gulf is one of the leaders and has a complete line of resins.

Applications include: medical and pharmaceutical products, packaging film, adhesive for laminates, and blending resin.

RUBBER HYDROCHLORIDE

Either bubble or tenter process is applicable to this material. Generally it is made into a tough, clear, heat-shrinkable wrapping in thicknesses as low as 0.5 mil. (Rubber hydrochloride is discussed in the Applications Chapter 6). Its importance is decreasing..

POLYBUTYLENE SHRINK FILMS (PB)

Oriented polybutylene film is noted for its toughness, strength, and "tailored" properties. The properties can be varied from moderately stiff with a modulus about 90,000 to 110,000 psi to a stiffer version of modulus 150,000 to 175,000 psi. The films are transparent, easily sealable, and shrinkable within well controlled limits.

When compared to vinyl shrink films, PB has less gloss and haze than PVC but has superior puncture resistance and general abuse resistance. (A test market in the mid-west of a polybutylene shopping bag showed that it outperformed anything available at comparable price). Table 37 lists the properties of PB film and compares them with the current competition.

The shrink properties of PB are close to irradiated polyethylene and are important in packaging materials that are easily crushed. The low creep properties of both PB and cross-linked PE film give very good shelf stability and integrity.

Plastic refuse bags are a well established product, and polybutene bags are superior in puncture resistance and retaining integrity in difficult environments (hospitals, restaurants etc.). They are particularly useful in

Table 37. Comparative Properties of Commercial Shrink Films

Property	PB	PVC	PP	XL'K PE
Tensile Strength, (psi × 10^4)	14-17	8-16	15-27	14-19
Shrink Tension, (psi)	300-600	150-300	300-600	250-1000
Shrink Temperature (°F)	185-400	150-300	220-330	200-280
Maximum Shrink, (%)	65-75	50-70	70-80	75-85
WVTR, [g/100m²/24 hrs/mil (72°F)]	0.22	1.91	0.14	0.12
Oxygen Permeability [cc(STP)-mil/m²/ 24 hrs (23°F)]	4600	855	1700	5670
Tensile Modulus (psi × 10^4)	10-12	5-15	20-25	6-9
Shrink Force (@95°C)				
Orientation	300-400	115-145	250-460	319
Contraction	100-150	115-145	250-460	325-10
Area Shrinkage @ 95°C (%)	30-35	35-49	10-15	20-80
@ 150°C (%)	80-85	70-75	80-84	90 @ 120°C
Optical: Haze (%)	1.3 ± .3	1 ± 0.5	1.1 ± .3	1.75 ± .25
Gloss (% at 45°)	83 ± 3	92 ± 3	87 ± 2	82 ± 2
Yield (in²/lb/mil)	30,000	22,000	30,000	30,000

Vol. 1 – "Science and Technology of Polymer films" by O. Sweeting, publ. J. Wiley, Interscience, pg. 286.

Table 38. Permeability of Polyolefin Films

Measurement	Polybutylene	LDPE
WVTR (g/mil/day-100 in²) @ 25°C 100% RH	.29	.30
Oxygen transmission [cc(STP)/mil/24 hrs/m²/atm]	6730	9400
Carbon dioxide transmission [cc(STP)/mil/24 hrs/ m²/atm)	3860	6500

Courtesy of Shell Chemical.

the area of preventing the spread of contamination in hospital refuse. Plastic refuse sacks are normally used in conjunction with a metal support stand. The lack of creep and puncture resistance make this an ideal system for disposal of contaminated waste.

Tests have been made with polybutene film as inflatable dunnage bags. This cargo stabilizing application appears to have been designed for the PB inflatable bag. When one compares the properties of the PB film with the current films used in this application one can see what can be expected especially when the PB film has a much improved impact resistance (Table 38).

NYLON FILMS (ON)

Polyamide (nylon) film has proven itself to be interesting and very versatile. The reason for its excellent potential is that it is a barrier to odors, flavors, oxygen, and petroleum products. Its ability to be thermoformed and still retain its mechanical properties or strength in all areas makes it an excellent packaging material. Nylon's poor MVTR (resistance to moisture transmission) restricts its uses to those areas where it is combined with other moisture barrier materials (i.e. coated or laminated).

Nylon film can be produced by conventional sheet extrusion followed by chill roll casting and blown film techniques.

Oriented nylon 6 can be produced by both blown film and tentering processes. (Nylon 6 is the polyamide of ε-amino caproic acid. All nylons are polyamides, that is, polymers formed by organic amines reacting with organic carboxylic acids.)

Some of the combinations (with other materials) are listed with the construction methods:

Combination	**Construction**
Nylon/PE and nylon/PP	Adhesive laminate or extrusion lamination
Nylon/PVDC	Extrusion coating

Coextrusion of nylon with other substrates is becoming the preferred route. This can be accomplished by both cast and blown film techniques (i.e. nylon/surlyn for oily and greasy foods and HDPE/Ionomer or ethylene acrylic acid/nylon/EVA combination to replace Pliofilm as a bag liner). Table 39 lists some of the nylon film applications and the important physical properties.

Metallized Oriented Nylon Film – Metallized, sealable biaxially oriented nylon film used in laminates (for example ON/adhesive/LDPE or copolymers) is an excellent oxygen barrier, with very good flex resistance. These properties of nylon make it an ideal candidate for gas flush packaging (coffee, aromatic products, and controlled atmosphere packaging of fresh meat are a few of the obvious applications).

Author's note

A double bubble process has been used in Japan for the manufacture of biaxially oriented nylon (ON) and polypropylene (OPP). This method uses an initial hot blown film (with larger than normal blow-up ratio) followed by a second "orienting" bubble with a blow-up ratio about two to one, that puts the shrink force into the film. (Courtesy of Cryovac Div'n, W. R. Grace).

Table 39. Properties of Oriented Nylon (ON) Required for Packaging Products

	Sausage Tray	Bacon	Cheese	Sub-primal Meat	Bake-in-Bags	Boil-in-Bags	Bread & Baked Goods	Pork, Poultry & Fish	Greasy & Oily Foods	Snack Food & Nuts	Coffee Packaging	Gas Flush Cheese	Cereal	Petroleum Products	Pet Food
Tensile strength				X	X					X	X	X			
Abrasion resistance	X	X	X	X	X			X	X	X		X	X	X	X
Impact resistance				X	X	X			X	X					X
High temp. resistance					X	X	X							X	
Low temp. resistance	X	X	XX	X	X	X	X	X	X						
Oxygen barrier	X	X	X	X	X	X	X	X	X	X	X	X	X		X
Odor & flavor barrier	X	X	X				X	X		X	X	X	X		X
Chemical & hydrocarbon barrier					X	X			X	X				X	
Thermoformability	X		X				X								
PVDC coating capability	X	X	X					X	X	X	X	X			
Heat sealing stability	X	X	X	X		X	X	X	X	X	X	X	X	X	X

Couresty of Modern Packaging and Package Engineering – Allied Chemical Data

Chapter 5

Properties of Heat-Shrinkable Films*

Generally, orientation improves tensile and impact strength, clarity, and flexibility at both ambient and low temperatures. The degree of shrink and the shrink energy come from the orientation process.

Gas-transmission rates and moisture-vapor-transmission rates for amorphous polymers, oriented and unoriented, appear to be nearly identical. Crystalline polymers show a significant reduction in moisture permeability when oriented. This difference is greatest at low degrees of crystallinity (10-15%) and becomes gradually less as the degree of crystallinity increases, until at 40-50% crystallinity no differences are discernible. The gas-transmission rates are largely dependent on the amorphous content, which outweighs any effect introduced by orientation.

On the other hand, orientation generally has a detrimental effect on elongation, ease of tear propagation and the sealability of film. The heat-sealing range is narrowed and the film may vary in properties with age. The maximum amount of shrink obtainable in a film can be calculated from the degree of stretch imparted to it during the orientation step. During shrinkage, the film again returns to its original condition.

Table 40 shows the effect of orientation on physical properties, specifically PET and PS.

*Appendix 13.2.

Table 40. Effect of Orientation on Physical Properties

	Tensile Strength, (psi)	Elongation at Break, (%)	Yield Stress, (psi)
Amorphous polyethylene-terephthalate			
Unstretched	6,000	500	6,000
Biaxially Stretched	25,000	130	~10,000
Uniaxially Stretched	>75,000	7	>75,000
Polystyrene			
Unstretched	*5,000-9,000*	*1.0-3.0*	*5,000-9,000*
Biaxially Stretched	*9,000-12,000*	*1.5-5.0*	~10,000
Uniaxially Stretched	15,000-18,000	6.0	10,000

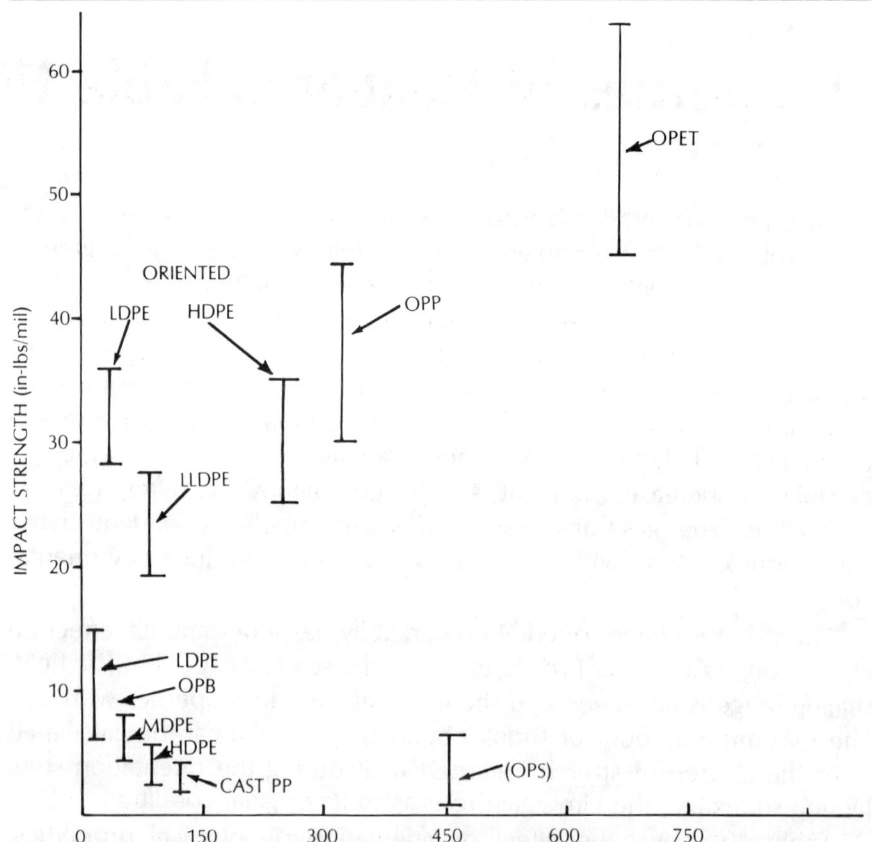

Figure 49. Secant modulus @ 1% elongation (psi × 10³) Includes LDPE (both low pressure and high pressure types. Sources: Encyclopedia Modern Plastics, Modern Packaging Ap 1966 and commercial data from Dow, Shell, Cryovac & UC&C.

Figure 49 describes the impact strength and Secant Modulus of the various shrink films.

DEGREE OF SHRINKAGE

Figure 50 shows that the maximum amount of shrink available in the films varies from 15-80 percent. The percentage of shrink increases with the temperature of shrinkage, so that the shrink obtained from a film can in theory be controlled in this manner. Unfortunately, this is very difficult to accomplish with most shrinkage techniques, since the percentage of shrink is usually determined by the object being packaged. The slope of the shrink versus temperature curve depends on polymer composition and manufacturing techniques. Films with steep shrink curves need close temperature control of the shrink equipment. For example, a temperature variation of ±5°F in a shrink tunnel at commercial shrink temperatures (i.e., 120-130°F) can cause the degree of shrink to vary theoretically up to 22 percent for polypropylene film, though only 6 percent for polyvinylchloride.

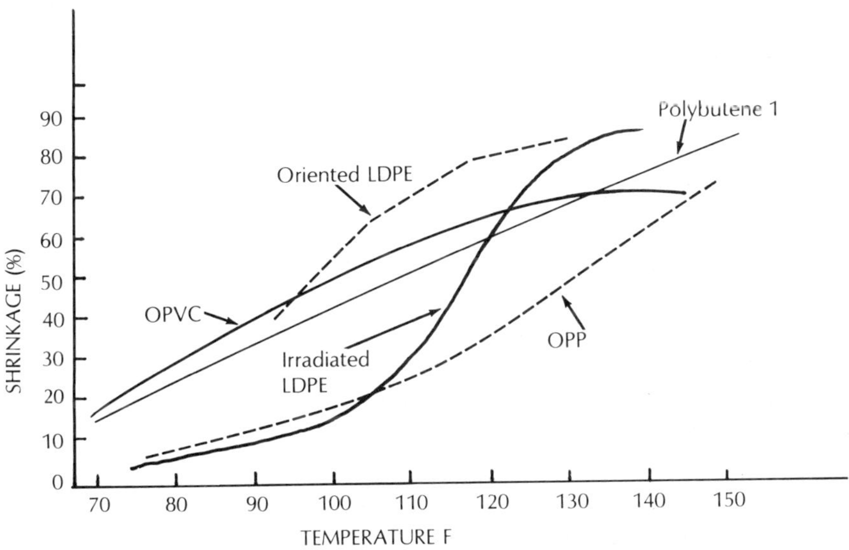

Figure 50. Shrinkage characteristics at various temperatures. Source: Modern Packaging Films, Edited by S. H. Pinnir Publ. Butterworth, London 1967.

Different applications need different amounts of shrink. Little is required for tightening a loosely wrapped package, while contour packaging requires a high degree of shrink. There are some special applications, such as sleeve wrapping, in which only longitudinal shrink is required.

Balanced orientation is particularly important for printed films, since uniform "area shrinkage" is essential to avoid distortion of the print during shrinkage. Even a balanced biaxially oriented film may not shrink evenly in both directions if the product is of very irregular shape, and in such cases it may be necessary to choose a print design which is not affected by such distortion.

SHRINK TENSION

Shrink tension is the stress which a film exerts when it is restrained from shrinking at elevated temperatures. It can be influenced by polymer properties and method of manufacture. Tension of 50 to 150 psi is always desirable in order to provide a tight package after shrinking. Higher tension (up to about 500 psi) is desirable where the film becomes a structural part of the package. With a shrink tension above 300 psi, care must be taken in limiting temperature and time to prevent crushing or distorting the package.

Figure 51 describes the shrink energy (tension) of oriented polypropylene, polyethylene, polybutene and PVC.

SHRINK TEMPERATURE

The temperature range of shrink is an important commercial consideration. Ability to shrink in boiling water is preferable, but films may be handled in equipment which provides hot air (100°-315°F).

Nevertheless, low shrink temperatures are preferable because of the simple equipment required and particularly for the packaging of heat-sensitive goods. Polypropylene film is deficient in this respect as well as in the shape of its shrink curve, though by special techniques it may be possible to improve the shape of the shrink/temperature curve and the optimum shrink temperature.

Exact control of temperature and air velocity is necessary to get satisfactory results when handling films which become very soft and weak near their melting point. Crosslinked polyethylene does not exhibit a drastic decrease in viscosity when it reaches its melting point, therefore its particular advantage lies in being able to withstand high temperature where it can easily recovery the maximum inherent shrinkage. (Ref. Figure 50).

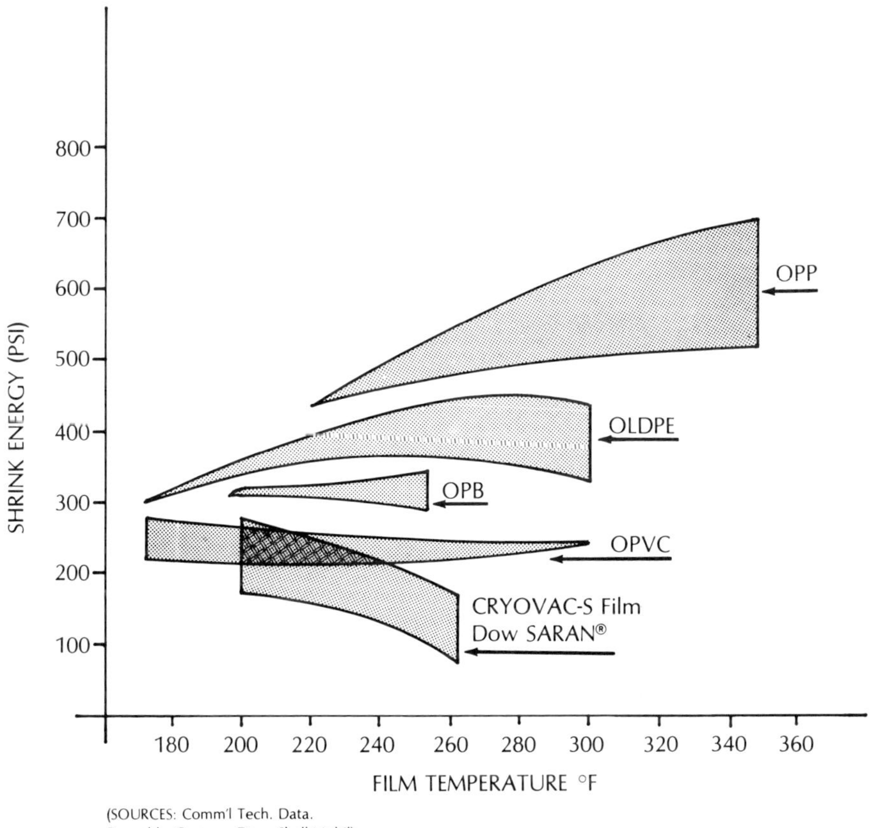

(SOURCES: Comm'l Tech. Data.
Reynolds, Cryovac, Dow, Shell-Mobil)

Figure 51. Shrinkage force (energy) at various temperatures.

SHRINK MECHANISM

All oriented films shrink at a temperature near their melting points or softening temperatures. The amount and temperature of shrinkage depend on the stretching and annealing history of the film.

AMORPHOUS POLYMERS

A typical biaxially oriented film of a non-crystallizable polymer, (for example, polystyrene) on gradual unrestrained heating, shows only a thermal expansion until the region of the second-order transition or glass-transition temperature, Tg, is reached. At temperatures above Tg, the material shrinks, the rate of shrinkage increasing with increasing temperature (Figure 52 and 53).

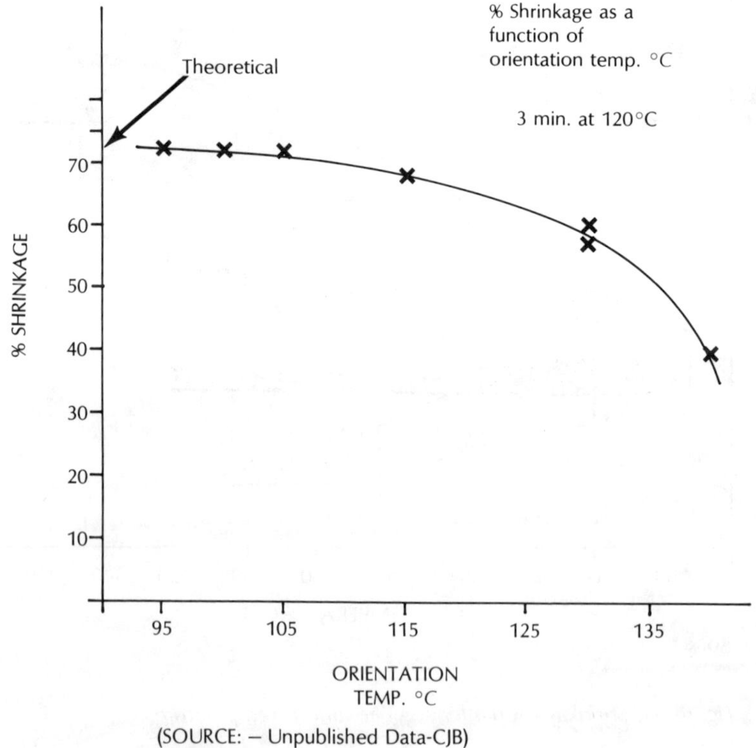

Figure 52. OPS - % Shrinkage recovered vs. orientation temperature.

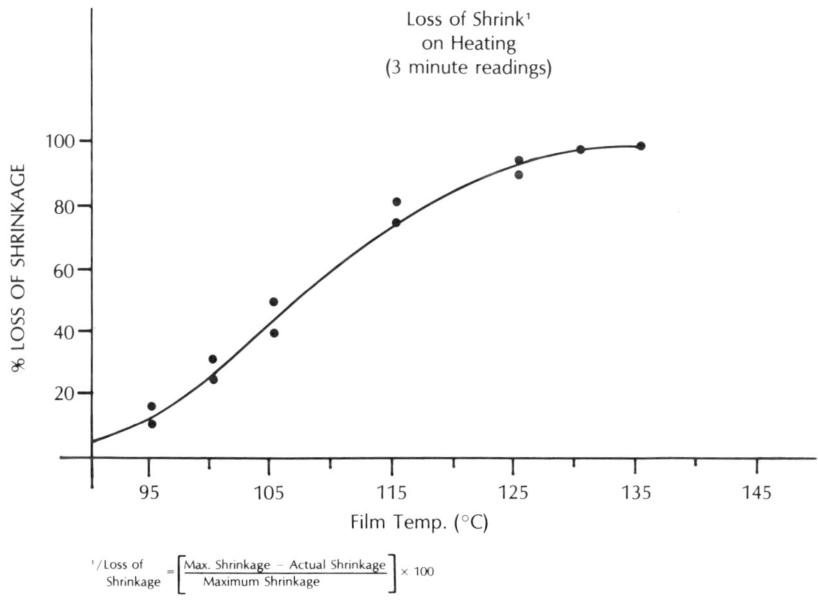

Figure 53. Stress relaxation and loss of shrink.

The shrinkage of oriented amorphous polymers is a relatively simple phenomenon. When the polymers are heated above their glass-transition temperatures, the frozen-in orientation begins to disappear as E_2 retracts. A very simple experiment clearly demonstrates this behavior. A streteched rubber band immersed in a dry ice-alcohol mixture, and thus cooled below its glass-transition temperature, retains its stretched dimension when the stress is removed. But when removed from the cold bath and warmed up to room temperature, the rubber band spontaneously reverts to its original dimension.

A highly oriented film can be shrunk almost to its original dimensions by heating, and the maximum amount of shrinkage of an oriented film is directly dependent upon its degree of orientation. For example, if the orientation is carried out at a low temperature, using a high stretching rate and a rapid quench, nearly all of the stretch put into the film is recoverable upon heating. If all other conditions remain constant and the orientation temperature is increased, less original stretch is recoverable due to relaxation (Figure 54).

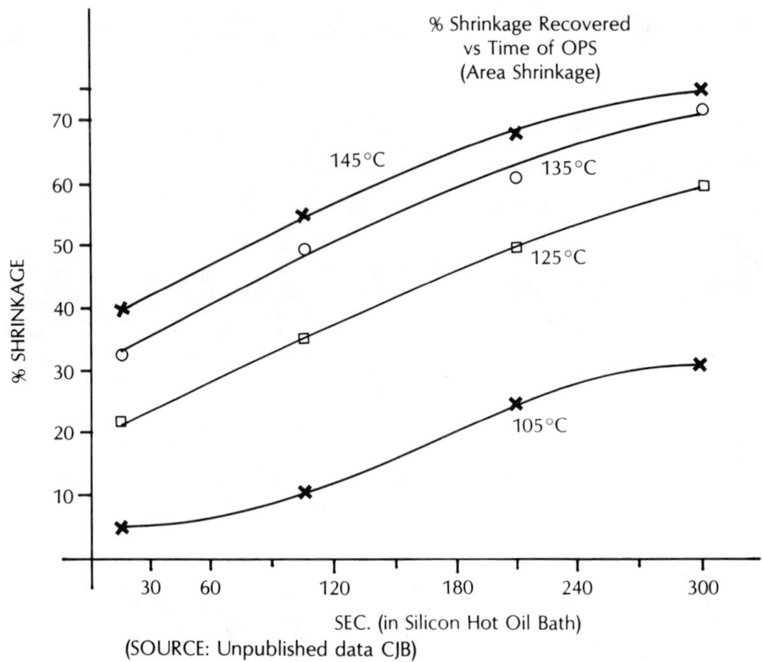

Figure 54. Shrinkage (Recovery vs. Time & Temp.) of oriented polystyrene.

The rate of shrinking increases with increasing temperature until the film is heated to a sufficiently high temperature for viscous flow to compete with retraction. Above this temperature, less than the theoretical maximum amount of shrinkage takes place.

When a sample is heated and restrained from shrinking, it loses orientation, but not as rapidly as if free to shrink. A restrained film has an orientation release stress when heated. That is, the material exerts a pull on the restraining clamps. The magnitude of this release stress is related to the original stretch conditions and gives a measure of orientation level, although it does not always correlate with the amount of shrinkage (Figure 55).

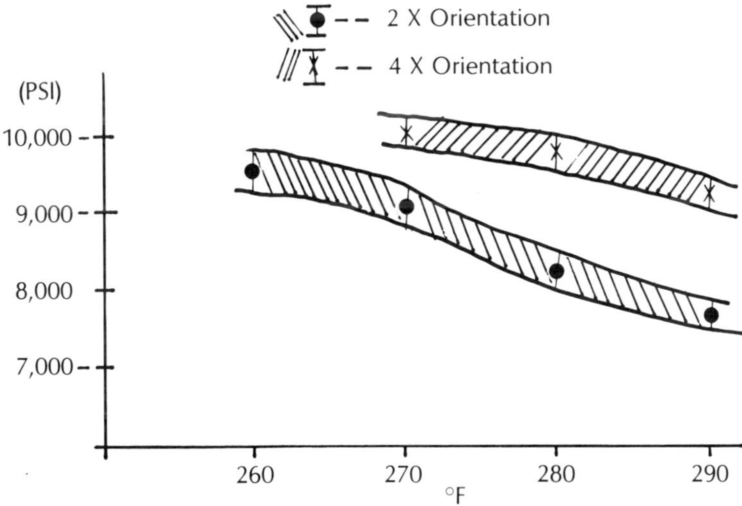

Figure 55. Effect of orientation temperature on tensile strength of polystyrene.

There is no known way to stabilize an oriented film of a noncrystallizable polymer against shrinkage above its glass-transition temperature.

CRYSTALLIZABLE POLYMERS

Biaxially oriented films of crystallizable polymers, such as polyethylene terephthalate or Saran™, in the amorphous state shrink in the same manner as those of noncrystallizable polymers. However, they differ from the latter in that they can be stabilized against gross shrinkage above their glass-transition temperatures. Thus, it is possible for two films of the same material and having the same degree of orientation to differ greatly in their shrinkage behavior at a given temperature. The film is stabilized by a heat treatment during which crystallization of the oriented polymer takes place. The stabilized film can still be made to shrink, but before it will do so to any appreciable extent it must be reheated to the heat-stabilization temperature. It is fortunate that this stabilization of crystallizable films is possible, since the glass-transition temperature of the majority of commercially important crystallizable polymers lie below 25°C (See Table 41).

Table 41. Glass-Transition Temperatures of Common Plastics

Polymer	Tg, °C	Crystalline Melting Point or Softening Point °C
Polyethylene		
High-Density	−120	135
Low-Density	− 25	98
Polyoxymethylene	− 50	185
Poly(vinyl fluoride)	− 20	195
Polypropylene	−20 to 0	175
Vinylidene Chloride Copolymer	− 17	195
Poly(vinyl acetate)	29	175
Poly(hexamethylene adipamide)	50	250
Poly(ethylene terephthalate)	69	255
Polystyrene	80-100(85)	105-110
Poly(vinyl chloride)		
No Plasticizer	105	212
15% Plasticizer	60	
Cellulose Acetate	70-120	300
Poly(methyl metacrylate)	105	160
Polytetrafluoroethylene	126	327
Polycarbonate	150	230

Ref: 13.1.2

Sources: U.S. Patent 3,231,643 to DuPont Jan. 66 and Sweeting's book (see Table 37).

ORIENTATION EFFECTS

The effects of molecular orientation on gas permeability vary widely depending on the particular polymer, additives, fillers, degree of adhesion between the filler, and resin. In general, orientation decreases permeability by 10 to 50% depending on the type of unfilled polymer, the degree of orientation, and the temperature of orientation. The following data shows the effect of orientation on permeability and Figure 56 compares the impact strength of cast polypropylene with oriented polypropylene.

Polymer	Degree of Orientation	Oxygen* Permeability
polypropylene	0	150
	300%	80
polystyrene	0	420
	300%	300
polyester	0	10
	300%	5
acrylonitrile/styrene	0	1.0
(70/30 copolymer)	300%	0.9

*(c.c/mil/100 sq. in./day/atm. at 73°F)

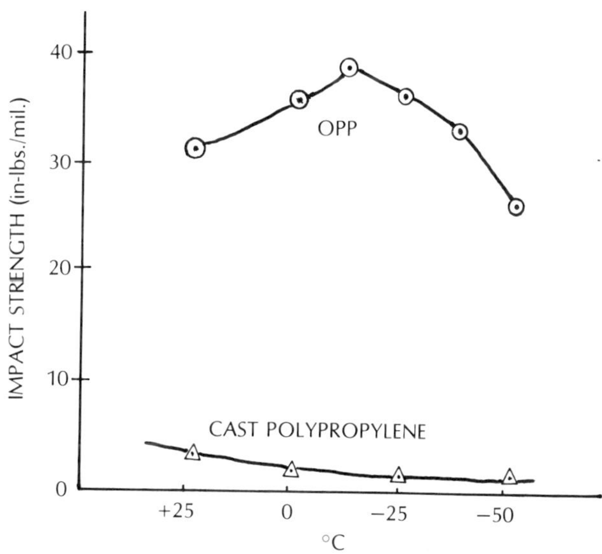

Figure 56. Impact strength of cast and oriented PP vs. temp. Courtesy of UC&C and from Modern Packaging April 1966, p. 208.

Chapter 6

Films, Laminates and Their Markets

GENERAL MARKETS

The use of conventional plastic films extends into such diverse areas as packaging, building construction, agriculture, health and medical supplies, clothing, and automotive applications.

Virtually every thermoplastic is used in film form. The majority of thermoplastic films are prepared by conventional extrusion, calendering, or solvent casting techniques. The extrusion method has two distinct advantages: the process requires relatively low capital investment and is capable of making thin film (less than 2 mil). The calendering method requires large capital investment, but is used successfully in large volume heavier film (e.g. PVC film greater than 2 mil). The solvent casting process is expensive and of less importance than the other two.

In general, there are several key markets for plastic films:

FOOD PACKAGING

In-store wrapping

Fresh meat – PVC (¾ mil), Saran™, coated PE, PP, EVAC

Fruit and vegetables – PVC, PS, PE, PP
Frozen meat – PE (cross-linked), Saran™
Snack meat – polyester, Saran™ coated composites
Pre-store wrapping
Processed meat – rigid PVC, Saran™, structured film and laminates
Tray – rigid vinyl, PS, foam styrene, ABS
Frozen food – PE, laminates, polyester
Bakery – PE bags, OPP, OPS, OPS trays, cellophane
Use in the home
Wall covering – vinyl, vinyl coated paper, coated OPP
Upholstery – vinyl, vinyl coated cloth or non-woven
Paneling – rigid vinyl laminated to wood. (Urea and phenolic resin impregnated paper)
Draperies – vinyl, resin coated cloth or non-woven
Table cloths – non-woven, PVC, (calendered, embossed)
Electronics
Magnetic Tape – polyester, PVC, laminates
Industrial Chemicals – PE, polybutene, PE/PIB composites
Clothing and accessories
Raincoats, dresses – PVC, Urethane, foam/film laminates
handbags, shoes
Health and medical
Oxygen tents – PVC
Sheeting – PE, rubber, PVC
Refuse – PE, polybutene
Tape, etc. – PVC, polyester, OPP
Food service – PE, PP, PS, foam/PS laminates

In other areas (i.e. industrial packaging and construction) film usage is restricted to bags, bags-in-box, over-wraps, and special functional composites (like dunnage bags, coatings, and specialty laminates to fit a specific property demand).

Polypropylene – Polypropylene is a crystalline polymer with a melting point of about 170°C. If PP film were not quenched rapidly after processing close to the melting point, spherulites would form and the product would be opaque and have poor impact properties. That is why film made by tubular techniques using air cooling, such as that used in PE production, is of less interest for packaging applications. Rapid quenching of blown film can be achieved by special techniques and flat film casting prevents formation of large spherulites, and transparency is maintained.

Most polypropylene film is marketed in thin film form (less than 2 mil) and is made by the biaxial orientation process. A heat-set film is preferred in order to minimize shrinkage on exposure to heat in printing or processing (e.g. coating, laminating or sealing). It is appropriate to compare the properties of polypropylene film with those of the two other major transparent films used in packaging applications, namely, polyethylene and cellulose films.

PVC Film – PVC film may be produced by casting, extrusion or calendering and the films coming out of these processes have very specific applications because of their physical properties. For example:

PLASTICIZED CALENDERED VINYL

ADVANTAGES: Moderate tensile, high elongation, tear resistant, water repellent, chemical resistant.

APPLICATION: Raincoats, toys, shades, drapes & curtains, covering.

RIGID VINYL

ADVANTAGES: Dimensional stability, high tensile, water and chemical resistant, flame retardant.

APPLICATION: Credit cards, wall coverings, laminates (wood and plastic).

CAST VINYL

ADVANTAGES: High gloss and clarity, chemically inert, MVTR low, high tensile and tear.

APPLICATION: Laminates, overlay film for food, furniture covers, skin packaging, in store stretch wrapping.

COPOLYMER FILM (SARAN™)

ADVANTAGES: Chemical resistant, fire resistant, MVTR very low, gas barrier very high.

APPLICATION: Barrier film (candy, cheese, meat, etc.).

Polyethylene Film— Polyethylene is classified as a crystalline polymer. The degree of crystallinity of commercially availabe polyethylene ranges between 55-98%, and the density varies correspondingly within 0.915-0.970 g/ml.

The crystallinity of polyethylene, and hence its density, has a major influence on its properties (Figure 57).

Crystallinity ↑	Elongation↓
M.P.↑.	Relative Impact Strength↓
Density↑	Gas Permeability↓
Tensile Strength↑	
Rigidity↑	

Figure 57. Effect of crystallinity on properties of polyethylene (at constant mol. wt.).

The main requirements of a packaging material are:

A. To protect and maintain the quality of the contents.
B. To enhance the esthetics of the product.

The properties that make polyethylene film a popular packaging medium are its low price, toughness, flexibility over a wide temperature range, pleasing appearance and softness, chemical inertness, relatively high oxygen and carbon dioxide permeabilities, and low water-vapour permeability.

On the debit side are its low softening point, poor resistance to greases, and difficult machine handling. The use of high-density film improves softening temperature, grease resistance, and stiffness but has poor clarity.

Applications of polyethylene film in the packaging industry are conveniently classified into two headings: food and industrial chemicals.

POLYETHYLENE

ADVANTAGES:	Low cost, good tear, low temperature properties, special shrink grades.
APPLICATION:	Bags; over-wrap; product packaging; agricultural, construction, drum liners; general purpose barrier; shrink wrap; fresh meat (stretch); cured meat (shrink); bakery products; milk; frozen food.

Polystyrene (OPS) – Polystyrene film with its clarity, crisp feel, and low moisture barrier is used for bagging lettuce. (At the present time packaging lettuce in the field is an automated operation. Baggers sit on a traveling harvester, equipped with washing, bagging, sealing, and shrink stations; the shrink-wrapped head of lettuce is protected from loss of moisture and gives a crisp feel without likelihood that the head will "die" and start to decompose.) This film is used also as windows in envelops and surface laminated to polystyrene foam for disposable plates, trays, and folding cartons.

Oriented polystyrene has also been applied in the electronic field and has been used to produce small capacitors. It has a dielectric constant of 2.5 (a little higher than HDPE) and a low loss angle. The resistivity is good (10^{19} ohms/cm). Since the breakdown voltage is low, it is used in low voltage capacitors, and small capacitors where the thin film is required.

OPS is suitable for thermoforming (i.e. trays for meat and produce and containers for confectionaries). The resulting containers are still brittle and can be cracked when bent. The polymer can be modified to improve its heat resistance and impact resistance. These are accomplished by copolymerization and increasing the crystallinity of the polymer or its stereo-regularity.

ORIENTED P.S. FILM (EXT-C/O)*

ADVANTAGES: Clarity, rigidity, FDA acceptance, low cost, very thin, high MVTR, high gas permeability.

APPLICATION: Packaging fresh produce, shrink wrap, disposable food service, stationary.

(*EXTRUDED–Cast/Oriented)

Nylon and Polyester – Other films that can be considered the workhorses of the packaging field are nylon film and polyester film. The bright areas for these films lie in shrink applications and special structures and laminates.

NYLON 6 FILMS (EXT-C/B)*

ADVANTAGES: Thermoformability, crack and abrasion resistant, good low temperature properties, FDA approved.

APPLICATION: Laminates – (N/PE)

Vacuum packs, processed meat and cheese, boil-in-pouch, bake-in-bag, hospital-medical, tape, bags, balloons.

(*EXTRUDED–Cast/Blown)

POLYESTER FILMS (C/O*)

ADVANTAGES: Tough, sterilizable, clear, chemical resistant, laminate base, low MVTR.

APPLICATION: Tape, metallized films, vacuum and gas packaging, shrink packaging, cured meat, boil-in-bag, thaw-in-pack, medical packs, electrical insulation.

(*Cast/Oriented)

Engineered Films – Engineered films such as polyallomer, polyimide, acrylics, and others have been developed for specific applications rather than a spectrum of areas. For example:

POLYALLOMER

ADVANTAGES: Optical clarity, exceptional strength.

APPLICATIONS: Tape, fibers, strapping, containers (most applications in high draw-down area).

POLYIMIDE

ADVANTAGES: Exceptional electrical properties, thermal stability, low temperature flexibility, fire resistant, radiation resistant.

APPLICATIONS: U V resistance, heat and light resistant.

Fire fighting gear.

ACRYLIC (KORAD)

ADVANTAGES: U V resistance, heat and light resistant.
APPLICATIONS: Decorative laminates for metal, wood etc. construction.

Perhaps one of the more interesting films from the standpoint of toughness is thermoplastic urethane film.

THERMOPLASTIC URETHANE FILM

ADVANTAGES: Tough, elastic.
APPLICATIONS: High altitude balloons, packaging sharp objects, oil pouches, blood bags, radiated food containers, military packaging.

Here is a system that can provide durable service under extreme operating conditions. The abrasion resistance of thermoplastic urethane is significantly higher than other conventional films and sheets. This property is not only useful in free film package design, but also imparts excellent wear characteristics to laminates when this film is the laminating base film. The impact properties of these materials are excellent because of their high elongation and puncture resistance. Thin film dart impact values are much greater than those of other conventional films of equivalent gauge. As one can conclude, from the list of applications above that these films are finding new specialized uses which command high performance.

Relationship Between Polymer and Film Properties – Polymer structure determines film properties to a major degree. Processing, on the other hand, also influences the behavior of the film. The major properties which affect film performance are molecular weight, degree of branching or crystallinity, molecular weight distribution, and the structure and location of the comonomer within the molecule. For example, as one increases crystallinity in a given system:

Modulus increases.
Softening point increases (this means sealing temperature, and shrink temperature increase).
Impact properties decrease, optical properties suffer, (optical properties can be improved by orientation).

As molecular weight of a given polymer system increases:

Sealing temperature increases.
Tensile, impact and stress crack resistant properties improve.
Low temperature properties improve.
Melt strength increases.
Optical properties suffer.

Applications of shrinkable films have been limited up to now to products such as foodstuffs, where the higher cost of the film and the process could be justified by improving storage quality. Lately, the market picture has been changing; shrink film is becoming more important in packaging containers and raw materials because alternatives (corrugated containers, paperboard, solid fiber; etc.) are becoming more expensive.

PRODUCE

Owing to perishability, the bulk of fresh fruit and vegetables is still sold unpackaged or packaged at retail or wholesale level in perforated polyethylene bags or nets. Long term transport of these materials must remain minimal to avoid bruising, and consequent decay of the loosely packed product. (Onions have a unique problem – packaging in any container that does not allow free flow of air through the package leads to the formation of fungus due to the high moisture content required for freshness.)

When packaging is required at the source or when an extended storage life is desired, the packaging film should have a high gas permeability and anti-fog properties.

Sleeve wrapping of tray packs with uniaxially or biaxially oriented shrinkable films secures satisfactory fixation of the produce despite the lateral opening in the package. Cross-linked polyethylene, oriented medium density polyethylene, plasticized PVC, and polypropylene films share this rapidly growing market.

The packaging of fresh vegetables and fruit provides the largest single use of printed polyethylene bags. The pre-packaging of potatoes, carrots, parsnips, beets, radishes etc., for supermarket sales has become an important aspect of food distribution. Because of the low water vapor permeability of polyethylene and oriented polypropylene, both films and bags are sometimes perforated to allow the product to "breathe." Permeation is the key to extending fresh produce shelf life.

Despite the recognized nutritional value of fresh produce a decreasing consumption of this type of food has occurred in the U.S. in the past 30 years. Frozen and canned produce has replaced much of the market.

The concept of controlled atmosphere (CA) for fresh produce rests on the fact that the living tissue of the product takes up oxygen and gives off carbon dioxide. A crude solution to this problem of generating a beneficial microclimate has been to use perforated films as noted here. Some of the newer films (Bordens Resinite™ RMF-61) have been so structured as to extend the shelf life of some specific foods, (bananas and tomatoes), to 30 days. Some of the highly plasticized PVC stretch films of the RMF-61 type have shown some encouraging results. There are two main objections to the use of PVC packaging films at the present time; high density (low yield) and environmental problems, (whether real or imaginary), both of which influence the market place. the environmental problems come from requests by the meat cutters union to reduce the odor of PVC stretch packaging during sealing in meat markets. This surfaced in 1973-74.

FRESH MEAT (Appendix 13.1.1)

Consumer inquiries have repeatedly shown that the housewife judges her supermarket by the quality of its meat, and it has long been established that packaging has a significant role in prolonging the shelf life of fresh meat. Fresh meat is very perishable and extending the shelf life of a valuable product and reducing loss of weight by dehydration or "freezer burn" has always been the objective of the meat packing industry.

Over 40 billion pounds of red meat is produced in the United States each year. Beef and pork alone account for approximately 35 billion pounds. It is estimated that 50% of this volume goes directly into retail outlets.

Following is an explanation of just how fresh meat packaging has evolved from hanging sides of beef wrapped in cheese cloth to controlled atmosphere packaging of primal and subprimal pieces in corrugated boxes that now dominate the wholesale market.

Before one can fully comprehend the reasoning behind the evolution of meat packaging concepts, one must understand the meaning of "fresh" meat.

Meat consists of approximately 16 to 18% protein, 18 to 20% fat and the balance is moisture with less than 1% minerals. Meat protein is a valuable dietary component and must be protected from biological attack and enzymatic degradation. Fat contributes the cooked meat flavor and texture and must be protected against oxidative degradation (rancidity).

Myoglobin, (the protein with iron porphyrin-like structures) is the major purple coloring element of meat. The meat color changes as meat "ages" or oxidizes.

For example, Myoglobin is the purple or ferrous porphyrin form, which upon oxidation turns to the ferric porphyrin form, or the red oxymyoglobin. (This is the "red color".) Further oxidation leads to the green porphyrin which is an irreversible change and cannot revert back to the "red" form.

Acidity of meat increases with age and certain proteins undergo enzymatic degradation into peptides and amino acids. This process, called "aging," develops the desirable texture and flavor. However, if this process is not controlled, the quality of the meat rapidly degrades and becomes unacceptable for human consumption. One can see that the control of oxygen and microorganisms is essential to prolonging shelf life. This is the main factor behind the packaging industry's interest in proper sanitation procedures, and procedures for packaging, processing and distributing meat, in other words, proper control of the environment that surrounds the product.

The best temperature for meat storage is just above freezing, but investigation of retail outlets has shown that it is rare that these temperatures are maintained.

Packaging films and concepts differ for wholesale and retail application. On the wholesale level, the federal regulations prohibiting the shipment of hanging carcasses between states has accelerated the use of centralized packing and distributing of sub-primal cuts. (For example, the use of heat-sealed and vacuum or gas flushed film wraps to reduce the loss of moisture and control the gaseous environment, in corrugated boxes.) This trend has reduced the cost of shipping, improved labor efficiency, and increased the yield. Packaging materials have been a decisive factor in this development.

The packaging material should have high oxygen permeability, to allow the formation of oxymyoglobin which is responsible for the desired purple "bloom" of fresh red meat. Cross-linked oriented polyethylene, polypropylene, plasticized PVC, and to a lesser extent, rubber hydrochloride and polystyrene, are satisfactory in this respect (polystyrene's moisture permeability is too high, and therefore allows meat to "dry out" in storage). This is not a preferred material. Oriented films have better sales appeal due to their "sparkle" and clarity. They can be used on hand-wrapping equipment and automatic machines. Shelf life can be extended beyond three days with these films if hygiene and refrigeration are kept under control.

Saran™ (vinyl chloride/vinylidene chloride copolymer) bags have very low gas permeability and are suitable for vacuum packaging where spoilage due to bacteria, surface oxidation and dehydration may be retarded up to 2 months. Vacuum packaging of fresh meat constitutes the key to central butchering, distribution and cost savings (less scrap, shipping weight, bulk packaging material purchases, etc.). This process of controlled atmosphere packaging (using gas mixtures for a flush prior to evacuating and sealing the bag containing the meat) will allow individual shops to be supplied with completely trimmed meat sections ready to be portioned and prepackaged for sale.

The high cost of 'back-room' cutting and packaging, plus the inherent health hazards, are contributing to the dramatic change in the manner in which meat will be distributed in the near future.

In fact, at the present time, the thrust to reduce the cost of fresh meat and extend fresh-meat shelf life is leading to the development of more efficient meat handling by establishing centralized meatcutting and packaging operations to eliminate the small supermarket cutting room.

New technology is emerging to meet this challenge 'to bring fresh meat to the consumer at a reduced cost.' All of the references that have been reviewed for this survey mention reduced cost and increased margin for the meat market but not reduced cost to the consumer. In fact, recent increases in red meat prices indicate the latter to be true. The development of systems to centralize meat cutting and packaging will inevitably ease the overhead cost burden of packers and supermarkets.

Standard Packaging Corp. has estimated that an average supermarket will save $33,000 by going to a centralized system. It follows that the large supermarket chains could save millions of dollars. In addition, the major meat packers can save money due to increased use of by-products, reduced shipping weight, more moisture retention, and improved quality. (An early Cryovac™ survey which indicated that by 1980 60% of retail beef and pork will be centrally cut, packaged and distributed has been verified).

At least four systems have been developed to extend the shelf life of meat and to prepare controlled atmosphere packaging — both requirements for a centralized system. If this trend continues (and based on latest market studies the emphasis is being accelerated) one can expect a major increase in packaging materials sold into this market.

Four systems of note are those of the Standard Packaging Corp., the Cryovac division of W. R. Grace, Union Carbide, and Auckerlund & Rausing. All of these systems concentrate on controlling the environment inside the meat package to extend the shelf life.

Standard Packaging's development consists of a lid made from a *polyester/PVDC/polyolefin* high barrier layer on a permeable substrate, rugged rigid base sheet and a machine that thermoforms the sheet to a tray. Standard's machine thermoforms the meat tray, gas flushes the interior, and creates a hermatically sealed retail package. (FlexlVac™: form/fill/gas flush/seal).

Cryovac developed a vacuum packaging machine plus a barrier bag of *EVAc/VCl$_2$-VCl/EVA* laminate specifically for vacuum packaging fresh meat. Vacuum/flush cycles cause the barrier bag to collapse around the meat eliminating the trapped air that deteriorates meat quality. The package is clipped, then 'heat shrunk' and boxed.

Union Carbide has the *Perflex*™ packaging system that is designed to bag and convey primal or sub-primal cuts of meat from the end of the processors boning table to a *conventional bag evacuator.*

Auckerland & Rausing have already reduced fresh meat packaging cost 6 to 8%, reduced labor cost by one third, and have improved sales 7 to 10% by centralized packaging and distribution to small self-service stores, cafeterias and dairies of IRMA, (a Danish corporation). The process vacuum forms trays from a roll of laminated sheet, injects oxygen and carbon dioxide and seals the barrier lid in place for a week of stable shelf life. As has been noted, the films used in the *wholesale* or centralized cutting and packaging areas are vacuum/gas flushed bags that range from high barrier Saran™ (oriented PVDC copolymers) to multilayer composites.

Some other specific examples of packaging films used in this application include:

American Can's non-shrink bags made from nylon/surlyn coextruded film and PVDC copolymer films (Maraflex™ HBSS)
Bordon PVC film and bags (Fab-Wrap™)
Cryovac Division of W. R. Grace EVAc/VCl-VCl$_2$/EVA (B-620)
Goodyear PVC (BW 9-Vita-film™)

These are but a few. More composites are waiting in the wings to be fully commercialized.

Shrink and stretch films having less demanding barrier properties are used in the *retail* trade where individual packaging of smaller units puts additional emphasis on esthetics and are less demanding in extending the shelf life for longer than two days, or possibly three. Some films used in this area are,

Bordon Chemical—PVC stretch films (resinite-RMF 61HY)
PVC shrink films (RMF-67 and 68H)
FMC————Cellophane (REO),
Nitrocellulose coated cellophane (MSBO-10)
PVC stretch film (MC)
PVC shrink films (HMS & RMS).

(NOTE: It is very difficult to compete with PVC stretch film in these applications, although with its rising cost and less yield some lower cost candidates are coming on the scene.)

It has been consistently pointed out (based on data from different sources) that the housewife or customer will continue to choose RED MEAT when it is available even at slightly higher cost.

Perhaps a more positive indication of what is to come is to look at the major paper companies and manufacturers of corrugated boxes. For example, International Paper has developed Blok-PakR™, a special packing system. The IPC system automatically erects and seals a one-piece corrugated carton. At least two Armour plants are using the system and each line is capable of erecting and sealing 180 loin boxes and 9 butt boxes per hour. One can see some "marriages" being formed between leading box makers and the makers of bags and sealing and controlled atmosphere packaging equipment.

FRESH FISH

Just as new marketing thrusts and packaging techniques opened up a large market in packaging sub-primal cuts of beef, so also there is an emerging thrust into the shipment of fresh fish. The packaging materials and the equipment are available.

At present the market is small but given the increasing demand for fish, the market could be a major user of specially constructed shrink films and bags.

Perhaps a close look at what a leader in the field is doing will clear this picture. Cryovac, a high technology, very astute marketing unit, makes a shrinkable ethylene vinylacetate – vinylacetate vinylidene chloride – Saran™ multiply bag. (EVA/PVDC-VAC/PVDC-PVC). This bag (specially designed for fresh fish packaging) has the properties to stabilize and preserve large sections of salmon so that an entire years supply can be packaged and distributed during the short salmon fishing season. The process consists essentially of processing the fish, vacuum packaging, and immediate freezing.

Icicle Seafoods® , together with Cryovac, developed a successful method of applying vacuum packaging technology that overcomes oxidation, dehydration and ice glazing.

The salmon is processed just prior to vacuum packaging and as soon as possible after the catch. The fresh processed salmon carcass (minus entrails) is then bagged for packaging. The bagged salmon is placed in a double vacuum chamber that allows it to inflate (by vacuum) and deflate (by pressure) to eliminate the oxygen within the bag. (When the oxygen (air) is forced out, the bag is completely formed around the carcass.) The bag is clipped to seal the package and finally the bagged fish is conveyed into a hot water spray shrink tunnel to finish the "tightening" (shrinking). This final operation presents a smooth, formed plastic shroud. The last step is to immediately freeze the finished package, box it and ship it.

I'm sure the reader will see the possibilities in this new approach. It begins to open up a large market that so far has been untapped.

(There is a similarity between this approach and the subprimal meat packaging that started a large market seven to eight years ago.)

CURED AND COOKED MEAT

Cured and cooked meat products, such as hams, bacon, frankfurters, bologna, and other sausages, are prepackaged under vacuum in shrinkable bags made of oxygen-impermeable materials, such as Saran™ and polyester. Vacuum and low oxygen permeability are indispensable in preventing bacterial growth and discoloration due to a light-induced oxidation of the nitrosomyoglobin. Colored films or film additives absorbing the critical light waves in the ultraviolet or blue range have found customer appeal in the United States.

FROZEN FOODS

Poultry has provided the oldest market for shrinkable films. The tight cling between the vapor-impervious packaging film and the product prevents freezer burn and pockets where ice may crystallize. The films used for this application are Saran™ and polyester bags which provide storage life of over one year. Where shorter storage periods are feasible, films such as cross-linked oriented polyethylene and plasticized PVC are satisfactory.

Frozen meat packaging at the retail level is accepted but is a very small part of the market at present. (Approximately 3.5% of the total.)

One can obtain stability in frozen beef for 18 months and approximately 12 months for fresh pork, if the temperature is maintained below zero to reduce enzymatic degradation.

Film wraps fulfill two essential requirements; they produce a barrier to loss of moisture (weight loss and therefore loss of sales dollars) and to oxidation or rancidity, in the case of fatty foods. Entire carcasses and sides of beef, lamb, and veal have been "protected" with polyethylene film wrap plus a non-woven sheet or stockinet to protect the material from punctures by sharp bones, etc.

Smaller pieces, like the sub-primal parts, require high barrier films to preserve meat integrity. These are laminates of oriented PVDC copolymers (Saran™) or nylon or polyester. Ionomer composites which have a given oxygen permeability, maintain the red oxymyoglobin color of fresh meat. These latter materials are sold by DuPont. PVC stretch films from Goodyear are also sold in this market.

Esthetics are an important factor in retail sales, and the absence of ice crystals (or opacity) on the inside surface of the package is important.

Freezing meat at the retail level is energy intensive and unproductive. (Centralized cutting, freezing, and distribution makes more "economic" sense.)

Daun and Gilbert very effectively point out that a semi-rigid tray (PVC) with a head space containing an adjusted oxygen/carbon dioxide balanced concentration could maintain meat color ten days, while fresh meat in air atmosphere lasted 6 days maximum.

The use of polyethylene film in wrapping of frozen chickens has created a large business in printed bags. Cryovac's radiation cross-linked oriented PE shrink film is the major material used in packing *frozen turkeys* and *large birds*.

Frozen vegetable packaging (peas, beans, corn, etc.) has become a sizeable outlet for 2 mil low density PE film. This application has developed largely through the introduction of form and fill machines.

High density PE film containing isobutylene to improve impact resistance is used mainly in "boil-in-bag applications, with some new type of polypropylene films coming onto the test market stage.

DAIRY PRODUCTS

All categories of cheese may be successfully packaged in shrinkable films, impermeable to oxygen and preventing moulds from growing on the surface.

In recent years, the curing of cheese in vinylidene chloride copolymer films and especially in vacuumized bags made of this polymer has become very popular, since this curing method prevents dehydration and rind formation. The rapid acceptance of film cured cheeses may be explained by changes in the eating habits of the consumers, favoring a softer, less salty flavor in cheeses.

The long storage life of cheeses prepackaged in shrinkable films based on polyvinyldichloride, allows for the prepackaging to be done by a central packing plant at wholesale or even at production level.

BAKERY GOODS

In 1970 one could say that the major drawback to the application of shrink films was the dominant use of cellophane film because of the lack of performance of shrink films on high speed packaging machinery. The marketing picture has rapidly changed, partially due to the perfection of "heat set" techniques with oriented polypropylene and due to the emergence of new specially coated films. Bakers are expected to ship $15 billion worth of bread, rolls, cakes, cookies, and pies this year. The growth rate is twice that of the GNP and with the latest figures could exceed a rate of 10% per year. However, in this dynamic market the impetus remains to develop more creative approaches in order to reduce RMC's and total manufacturing overhead. The bakery industry is continuing the trend to replace expensive cellophane with OPP.

At least 40 new packaging machines have been developed and tested for specific products within the industry in the past three years to reduce cost and improve sales appeal.

In the film area, some of the cost effective developments are:

American Can Co.'s pre-made *metallized polyethylene bags* that cost approximately 6% less than the foil/paper bags for French bread. A new packaging system for bread dough has also been developed that is a laminate of parchment (paper that can resist baking temperatures and is replacing foil and glassine).

Coextrusion of polypropylene homopolymers and copolymes are continuing their penetration of the cookie overwrap market as a replacement for imported cellophane. (In some areas oriented polystyrene overwrap is beginning to be used with the oriented polystyrene vacuum formed tray.)

Oriented polypropylene is replacing cellophane since it is currently about 40 to 50% less than the price of imported cellophane. This is being reflected in ITT's increasing use of OPP in snackcake packaging and the belief that ITT will replace PE film used in bagging bread with OPP.

A unique laminate being tested by Olin is a highly transparent cellophane reinforced with OPP. This laminate or composite is extending the shelf life and freshness of health food items to ninety days.

Perhaps a hint of what the future may hold is a development in France that guarantees the freshness and shelf life of bread for 3 months without preservative additives. The film is transparent and printed. It consists of a *Nylon 6/polypropylene laminate*. Bread is packed, nitrogen flushed, sealed and sterilized.

There is little doubt that polypropylene films will continue their penetration of baked goods packaging at the current rate of 7 to 10%/year.

BAGS AND SACKS

In 1981, 200 million pounds of high density polyethylene was converted into bags and sacks. These applications included merchandise bags, "T-Shirt" sacks, trash bags and bag-in-box applications for cereals, dry mixes, cookies, crackers. Another 12 million pounds went into a new application, "Fish and Deli" bags for in-store point-of-sale packaging. At the same time, 535 million pounds of low density polyethylene went into such markets as carry-out, merchandise, self-service, garment, and counter bags. Almost 25% of the application tonnage was in heavy duty sacks (Table 42).

The trash bag market is reported to be 525,000 metric ton/year. A list

Table 42. Bag & Sack Markets for Polyethylene Resins (In Million Pounds/Year)[a]

HDPE	1980	1981	LDPE/LLDPE	1980	1981
Merchandise Bags	66	143	Carry-Out Bags[c]	24.2	40
T-Shirt Sacks	–	2.2	Merchandise Bags	48	66
Trash Bags[b]	–	8.8	Self-service Bags	48	48
"Bag in Box" & Food Bags	55	48-53			
Fish & Deli Bags	11	12	Heavy Duty Sacks	99	103
			Rack & Counter Bags	158	174
			Trash bags[b]	700	800

Note:
[a]These figures represent actual pounds produced.
[b]Annual Trash Bag production capacity has been estimated to be 525,000 metric tons. This figure is based on actual extrusion capacity, installed and operating in February 1982.
[c]The market potential for grocery sacks has been estimated to be 450,000 annual metric tons. However, this figure is based on Kraft paper grocery sacks. Since polyethylene grocery sacks are lighter the market potential will be less.
Courtesy of Modern Packaging and Plastics World.

of the major converting facilities are contained in Table 43. These producers account for almost 95% of the U.S. trash bag production.*

There are three major resin competitors for the bag and sack market. They are LDPE, LLDPE and HMW-HDPE. A summary of the constructions and a prediction of the product mix in the next seasonal year is contained in Table 44.

The grocery sack market has long attracted plastic film and bag producers and the advent of HMW-HDPE in the U.S. and lower cost LLDPE may make this market more accessible. So far PE holds a mere 2% of this market dominated by Kraft Paper.

As noted previously three polymer candidates are believed to be in the forefront:

> *LLDPE blends* are being marketed in the form of a "T-Shirt" sack. Formulations ranging from blends of 30% to 50% LLDPE are being sold. The LLDPE allows the producer to make a thinner bag without loss of properties. Examples described in Modern Plastics, December 1981 issue point out that 1.1 to 1.5 mils of LLDPE blends are replacing 1.7 to 2.0 mil LDPE sacks. Mobil is marketing a 1.1 mil (LLDPE/LDPE blend) grocery sack. In 1979 to 1981, Union Carbide and others began introducing 1.1 to 1.35 mil *LLDPE (100%)* – grocery sacks (1/6 barrel size). Attempts are underway to downgauge further to 1.0 mil.

*See Appendix 13.1.1 #2.

Table 43. Trash Bag Production Facilities (Source: Modern Plastics January 1982)

Presto Products	*Bemis*
Appleton, WI	Flemington, N.J.
Union Carbide	*Capitol Poly Bag*
Rogers, AR	Cols. OH
Cartersville, GA	*Chicago Transparent*
E. Hartford, CT	Chicago, IL
U.S. Ind. Poly Tech.	*Exxon Film*
Bloomington, MN	Marietta, GA
American Cello	*Fortune Plastics*
Louisville, KY	Lebanon, TN
Armin Poly Film	*Frontier Bag*
Elizabeth, N.J.	Kansas City, MO
Jersey City, N.J.	*General Plastics*
Bes Pak	Miami, FL
Montgomery, AL	*Georgia Pacific*
Monsanto	Hamlet, N.C.
Anaheim, CA	*Glopak*
Norchem	Newark, N.J.
Mankato, MN	*Heritage Bag*
AEP Ind.	Dallas, TX
Mathews, N.C.	*Mohawk Plastic*
Fortune Plastics	Benardston, MT
Old Saybrook, CT	*PCL Packaging*
International Paper	Remington, VA
Jackson, TN	*Poly Print Pack*
Arrow Ind.	Plainview, N.Y.
Corrolton, TX	*Safelon*
Atlantic Plastic	Yonkers, N.Y.
Irvington, N.J.	

Sources: Plastics World Jan. 1982 and Modern Plastics Jan. 1982.

(The incentive is coming from some thin HMW-HDPE sacks being marketed that are 0.7 to 0.8 mil.)

HMW-HDPE grocery sacks are knwn in Europe and bag producers in the U.S. are preparing to produce HMW-HDPE bags. Himolene, CA is supplying HMW-HDPE trash bags in 3 sizes with significant thinner, lighter weight construction than LDPE. Phillips has resin technology ready to commercialize. Soltex has 2 new HMW-HDPE resins to compete with Union Carbide materials.

Paper companies are being involved. One cannot expect the forest products industry to sit back when their paper market is threatened.

Table 44. Product Mix in Bag and Sack Markets

Market	Product Mix	Estimated Product Mix
Trash Bags	Coextruded Film LDPE/LLDPE/LDPE LDPE/LLDPE BLends LDPE	LLDPE/LDPE 70/30 in1985
Millinery Bags	LDPE HDPE LLDPE (Blends)	LLDPE - 70% in 1984 LDPE – 20% in 1984 HDPE – 10% in 1984
Grocery Bags	Kraft Paper HMW-HDPE LDPE LLDPE Blends with LDPE & HDPE	HDPE (HMW) plus LLDPE blends could achieve 3% penetration by 1983.
Produce Bags	LDPE LDPE/EVA LLDPE Blends	LLDPE – 70-75% LDPE – 20-30% (in 1984)
Frozen Food Bags	Coextruded EVA/LLDPE LLDPE Blends & Coext. Films	LLDPE LDPE 50/50 in 1984
Heavy Wall Bags	LDPE HDPE (HMW) LLDPE Coextruded LLDPE/LDPE	LLDPE – 50-60% by 1986 LDPE – 40-60% in 1986 HMW–HDPE – 5% in 1985
Fish & Deli Bags	HDPE Coated Paper	HDPE will penetrate at least 30% of the market in 1984

Sources: Plastics World Dec. 1982 pg. 69-72; Plastics World June 1982 "Grocery Socks;" Modern Plastics Dec. 1982 pg. 42-45.

Making plastic bags is not a radical departure for the paper companies.

Table 45 lists eight forest products company's plastic film converting operations. The industry is well positioned to go into the plastic grocery sack business. In addition to these inplace facilities (which make trash bags now) other actions are planned. For example:

St. Regis is commercializing a LLDPE blend grocery sack.
Georgia Pacific is now sellng LLDPE sacks.
Trinity Bag & Paper acquired a subsidiary of a Canadian PE sack producer.

A study by Windmoeller & Hoelscher lists 15 U.S. participants in the

Table 45. Forest Products Corporations – Plastic Film Facilities (Plastics World, January 1982)

American Can (Est. Cap. - 30 million pounds/year)		
Des Moines, IA	LDPE	Blown Film
Neenah, WI	LDPE	Blown Film & Ext. Coating
Bemis Bag (Est Cap. - 80 to 130 million pounds/year)		
Terre Haute, IN	LDPE/HDPE	Blown Film
New London, WI	LDPE/PP	Ext. Coating & Blown Film
Flemington, N.J.	LDPE	Blown film
St. Louis, MO	LDPE/HDPE	Blown Film
Union City, CA	LDPE	Blown Film
Crown Zellerbach (Est. Cap - 110 to 150 million pounds/year)		
Orange, TX	LDPE/PP	Cast Film
Greensburg, IN	LDPE/HDPE	Blown Film
New Castle, Del.	PP	Cast Film
San Leandro, CA	LDPE	Ext. Coating
Georgia Pacific (Est. Cap - 35 million pounds/year)		
Crosset, Ark.	LDPE	Ext. Coating
Hamlet, N.C.	LDPE	Blown Film (Bags)
International Paper (Est. Cap - 25 million pounds/year trash bags)		
Jackson, TN	LDPE	Blown Film
Ludlow (Est. Cap - 30 to 40 million pounds/year)		
Homer, CA	LDPE/HDPE	Ext. Coating
Newark, OH	LDPE	Blown Film
St. Regis (Est. Cap - 110 to 160 million pounds/year)		
Birmingham, AL	LDPE, HDPE	Blown Cast Film
Dallas, TX	LDPE	Cast Film
Dallas, TX	LDPE/HDPE	Cast Film
Union City, CA	LDPE	Blown Film
Louisville, KY	LDPE	Blown Film (shrink film)
Union Camp (est. Cap - 25 million pounds/year)		
Providence, R.I.	LDPE	Blown Film
Tomah, WI	LDPE	Blown Film

Source: Plastics World January 1982 pg. 38-44.

grocery sack business (Table 46) as of December 1981.

Perhaps one of the best kept secrets is the exact cost of a 1/6 barrel Kraft Paper Sack and how much added cost would be involved if the Kraft Paper Bag Industry started resin reinforcing thin paper and developing a rigid-when-wet paper bag. The cost comparison used in Modern Plastics assumes a selling price of 2.9¢ to 3.3¢/sack and then calcualtes the cost of producing such a sack using LDPE, LLDPE/LDPE Blends, LLDPE and HMW-HDPE as a percentage of sack selling prices (2.9¢ to 3.3¢/sack). The numbers associated with these systems are very enlightening:

Material	Bag Thickness	Sack Price % of 1/6 Barrel
LDPE	1.75 mil	54 to 69%
	2.0 mil	62 to 79%
LLDPE/LDPE Blend (30/70)	1.2 mil	37 to 48%
LLDPE	1.0 mil	31 to 43%
HMW-HDPE	0.7 to 0.8 mil	32 to 43%

Based on these figures and the resin properties, it appears that HMW-HDPE may have the edge.

Table 46. Grocery Sack Procers (Source: Modern Plastics – December 1981)

Celloplant, U.S.A., N.J.
Georgia Pacific, CT
Hilex Poly, CA
Himolene, CA
Mobil Chemical, N.Y.
Poly Print Packaging, N.Y.
Tort Poly, N.H.
Rex Rosenlew, N.C.
St. Regis, CA
Sidney Bag, FL
Sonoco, Prod., MA
Tag Poly Bag, GA
Trinity Bag & Paper, VA
Buffpack,
Union Carbide, CT
Western Kraft, OR

Source: Modern Plastics Dec. 1981.

BOXED GOODS

Boxes containing toys, games, housewares and foods can be over-wrapped with heat shrinkable films such as oriented polypropylene, PVC and cross-linked oriented polyethylene. By engineering the shrink tension and properly controlling it, the finished package can receive an attractive appearance, the covering being unwrinkled and glossy for extended periods of time. Trim-sealing is the more prevalent method of sealing over-wraps since it eliminates both sidefolds or underfolds.

BUNDLING

The bundling of identical items such as cans and bottles, stationary products, etc., and the attachment of sales stimulating gifts are an established practice of sales promotion, and provides a rapidly growing application for stretch and shrink films, especially those having sufficient shrink tension and impact resistance combined with good sealability and low cost. The lowest cost films are oriented low-density polyethylene and plasticized PVC.

There is a trend toward central warehousing and automatic order filling. In these cases, where very high volume and palletized mixed orders are sent to various locations, stretch or shrink film will become a more important part of the system. Bundling with film is less expensive than large corrugated containers and steel strapping. With delicate items such as toilet tissue and paper toweling, diapers, napkins, etc. one can expect the large corrugated container to remain firmly entrenched because of the protection, load stability, and sales appeal of a built in display with appropriate graphics.

Chapter 7

Shrink Vs. Stretch Films (A Problem of Cost/Effectiveness and Energy)

Lower capital costs, less energy and the use of film that maintains high shrink tension under load with 35% reduction in cost are the key elements that are beginning to appear in the development of a major stretch film market. Several major trends are worth noting.

Tyson Foods switched from shrink wrapped chickens and chicken parts (on a polystyrene foam tray) to stretch PVC film. According to Tyson PVC film holds its tightness during long distance shipments. Production rates have doubled and film costs are down approximately 35% by their reports. (There is still some resistance to using PVC stretch film in some supermarkets. The fumes that are generated are usually due to improper sealing temperatures and a low level of sealing equipment upkeep.)

There is no doubt that stretch film, especially in the material handling area (*pallet wrap, bundling,* etc.), will continue to grow. In many of these areas the high performance of true shrink films may not be required.

Companies such as American Can (Omni film), Lantech Inc. F M C, Northern Petrochemical, and others are very active in the field.* (Experience indicates that the high margin, high performance areas will remain in the area of oriented films and structured composites, while the stretch films will become commodities and subject to the same price erosion noted in other areas.)

SHRINK VS. STRETCH

Today and tomorrow, because of its economic advantages, stretch film is expected to develop rapidly for most pallet and bundling applications. (The main advantages center on less capital required for wrapping a unit under tension and no heat energy required to shrink the film around the unit. Stretch film is simply wrapped or wound around the load to be constrained and sealed. (If proper "cling" properties are utilized heat sealing is not required.) The elastic memory properties are built into the film by extrusion techniques and resin structure.

The preceding paragraph capsulizes information found in many trade and technical publications.*

Let's consider some of the aspects and alternatives.

Energy – There is no doubt that energy requirements for stretching a film over or around a unit is less than using heat to shrink the over-wrap. But one must consider the total energy required from film roll to packaged unit.

Versatility – Shrink film conforms to irregular shapes. If one has mixed loads or is a major distribution center where mixed pallet loads and odd loads are shipped, this must be considered. Many food items are not geometrically regular.

Creep – There are many cases where stretch film has been over-extended and the packaging film has lost its tension. (Creep resistance is highest with cross-linked structures and increases with the molecular weight of the polymer film.)

Equipment – Stretch equipment is less expensive than shrink tunnels and accessories. However, one must consider the cost of labor, the throughput, and the cost of draping a pallet with a shrinkable shroud vs. wrapping on the machine. Published accounts indicate that machine cost and labor costs of each system are at a stand-off.

*See Appendix 13.1.1.

Shrink-Packaging – Shrink-packaging involves two steps: wrapping the unit with a sleeve or a full wrap and applying heat to shrink or tighten the film around the unit. This means that a wrapper and a shrink tunnel are involved in the simple operations.

Wrappers – The wrapper can be a simple impulse radiant L-type sealer. This type of sealer operates with a center folded film or roll stock "folded-in-line." The L-shaped sealer produces fully sealed packages. (The side seal of one package is formed at the same time the preceeding package's seal is completed.)

The fully automatic versions can seal 30-40 packages a minute. Semi-automatic sealers can handle packages as large as 30 × 70 inches.

Another trim-seal of "full-wrap" machine uses two rolls of film. The product is conveyed through the plastic curtain and the film webs (sealed by the end seal of the previous package) cover the product and are sealed as the product moves past the rolls. This end seal forms the sleeve around the product. The next station finishes the end seals.

These machines can wrap 25-35 units/minute intermittently and with some of the new units, 40-50/minute with continuous motion sealing and trimming.

Units for wrapping canned goods on a corrugated tray are reported to produce 100 shrink wrapped trays/minute.

Shrink Tunnel – The shrink tunnel is a recirculating hot air chamber, through which the loosely wrapped product is conveyed and subjected to 2 to 3 seconds of heat to shrink the film around the package (back to its original size). In this short time, the product is unaffected. (There are many varieties of conveyors, with engineered air flow to conserve energy, etc.)

Stretch Wrapping – Stretch film wraping involves a simple process of stretching a film around a container or unit and allowing the built-in elastic recovery properties of the film to constrain the unit. Some films are "hi-cling" films that require little or no heat sealing. Stretch wrapping can be the "spiral wound" type of wrapping or a wide continuous "1-pass" film. (The resins commonly used in this application are PVC, LDPE and LDPE/EVA polymers and copolymers where stretch properties are built into the film by the resin formulation and structure.)

Stretch film packaging techniques are most applicable for cube or rectangular shaped loads. Systems require less energy (no shrink tunnels). Data published in the July 1979 issue of Package Engineering indicate that research directed to evaluating shrink film has shown it to be "more forgiving and not as subject to variations due to changes in film, plant environment or shrinkability."

Several types of equipment are common in stretch film packaging.

Rotary-Full Web is a system that uses film width slightly longer than the height of the load. A complete wrap is made per revolution.

Rotary-Spiral Wrap uses narrower webs and as the load rotates on a platform the web is powered up and down to achieve a spiral overlap.

Dual-Roll Pass Through is as noted previously. Two rolls, positioned vertically and sealed together from the previous load, surround the load as it is conveyed through the film. The film is then stretched around the rear of the load, (stretching and sealing the film). These are high volume pass through systems capable of 100 loads/hr.

Companies such as Brockway Glass and Miller Brewing Co. are very active in pallet wrapping studies. Millier is reported to have shipped 160 rail cars of beer up to 2000 miles/shipment with minimal damage (which *could* indicate over packaging.

Zone Wrapping is a new technique used to reduce film cost and to allow pallets to "breathe" when required as with produce.

Annual increases of 30% are forecast for the stretch film business in the next 2 to 3 years.

With this background let's review some of the points to be considered in choosing whether to go shrink or stretch, and the precautionary measures.

Advantages and Disadvantages

Advantages are:

(1) Low process costs.

(2) Low fuel or energy requirements and

(3) Less capital investment per unit output.

Disadvantages are:

(1) Correction of tension during wrapping is essential. (Too much tension can crush the shipping cases or cause the film to lose its elastic properties (relax).

(2) Four sided coverage is provided while shrink covers can provide full protection (this means protection from the weather is lacking).

(3) Stretch film appears to have an advantage in large rectangular or uniform loads. However, stretch film technology cannot match the efficiency, cost and performance of high

speed over-wrapping machines for small packages, and stretch film is not applicable to irregularly shaped products.

(4) Based on several recent articles on stretch film application, it now appears that there may be some loss in load stability where the distribution time is extended. (Stretch films do exhibit "stress relaxation" with time.)

If the film is "over stretched" it loses elastic recovery. Young* has noted that properties of some stretch films diminish at 25% elongation, but his data indicate that relaxation of 5% within 24 hrs. for some systems is not unusual. As a result, load stability must be monitored.

Shrink Film Packaging – Advantages:

(1) Shrink over-wrapping of small packages is excellent for esthetics, protection, and cost effectiveness.

(2) Shrink packaging can work with almost any product shape and can be removed easily.

(3) The shrinkable bag approach to pallet wrapping is the simplest approach. (Here the shrink tunnel, hot air, steam, etc., cost could affect the spiral winding cost.)

(Ref. Appendix 13.1.1)

Chapter 8

Non-Packaging Applications

CONSTRUCTION

Late in 1978 a Solvay report noted some of the latest developments in Europe. These are now starting to appear in the States. According to the report, "Latest developments in Europe directed to improving the UV resistance and the impact strength of rigid PVC sheet promise to give a boost to PVC sheet roofing material. Rigid PVC sheet has been widely accepted by the building industry as a decorative exterior siding and as a basic material to reduce maintenence. In the mid-seventies the use of PVC as a roofing material was almost non-existent." Lately, test marketing and field testing in the United States has illustrated what has been expected due to development research in the late sixties, namely, that molecular orientation improves PVC's rigidity, toughness, weatherability, and low temperature performance. For example, Plastic Engineering published some relative values of PVC oriented for use in this application (Table 47).

INDUSTRIAL CHEMICALS

The use of heavy-duty plastic bags for fertilizer and the bulk packaging

Table 47. Mechanical Properties of Oriented and Regular PVC Sheet (Source: Plastics Engineering)

Property	Oriented	Regular
Stress at Yield (psi)	8,100	7,300
Ultimate Tensile St. (psi)	13,000	7,300
Tensile modulus (psi @ 1% El)	450,000	425,000
Elongation at Break (%)	110	175
Flexural Strength (psi)	14,200	13,800
Flexural Modulus (psi)	430,000	400,000
Tensile Impact (ft. lbs/in^2)		
70°F	840	330
32°F (0°C)	840	130
−4°F	840	50
−40°F	830	0
−76°F	800	0

of other chemicals is a development which has expanded rapidly in the past ten years, and polyethylene has the major share of the market. (Polybutylene bags have been test marketed and shown to provide more protection at the same thickness and size, however, the cost of polybutylene is not competitive with the standard PE shipping bag. Polyethylene bags make up 25% of the total weight of PE film sold for packaging.

Typical products packed in these bags are fertilizer, peat, plastic granules, asbestos, fillers, animal feed, and industrial chemicals such as ammonium nitrate, cellulose acetate, ferrous sulfate, and nitrocellulose.

Chapter 9

Coextrusion

The ability to develop the lowest cost functional film is of prime importance to a successful and profitable program. This becomes more significant with the advent of new lower cost resins (like LLDPE) and the potential for continued resin price increases. Coextrusion is a technology that combines resins in an effective and cost efficient manner, and fills a unique position in the conversion of resins into high performance films. The ability to combine 6 or 7 layers in a final structure and to combine such resins as PE, PP, EP, N. EVA, EVAL, PS: with barrier and heat seal coatings gives one an idea of the potential and versatility of this technology.

TECHNOLOGY

The key to quality in coextrusion is the ability to design the proper feedblocks and dies to accurately control the rate of extrusion and laminar flow of the various components. The object is to minimize the rheological differences of the various resins and to match the surface velocities of the components or the resin's flow from extruder, through feedblock and die into a final laminar structure.*

*Composite Container Technology – Modern Plastics – February 81 page 37.

Surface velocities of polymers with wide variations in mass, cause turbulence and result in non-uniform structures wherein the various laminar layers are non-uniform and in some cases incompatible. With similar materials (for example ethylene propylene copolymers, alone or with polystyrene: LDPE/LLDPE, HIPS/PS) the turbulence phenomenon is minimized. In many cases an adhesive layer "smooths out" the inconsistencies.

Coextrusion of nylon and polyethylene is particularly difficult with variations in melt viscosity greater than two.

One must match extrusion rate and laminar flow to control laminar layer thickness. The controlling factor is the extrusion rate of the mainstream polymer. The other resins must be balanced to this rate to produce uniform layers with good adhesion.

Flat die extrusion can combine up to nine polymer streams into a single channel "coat hanger type" die using the feedblock approach. Feedblocks (based on Dow and Chemical Welex/Cosden) generally restrict the choice of materials to those of similar rheology; adhesive layers (i.e. special resins that allow slippage and increase compatibility) are used to combine dissimilar materials such as nylon/polyethylene.

Companies like Composite Container have specifically designed feedblocks that contain "backflow" reservoirs in the flow channel.* Their technique reportedly allows one to match melt velocities with less difficulty. (Basically the feedblock is not the ideal solution for combining dissimilar materials such as polyolefins, nylon, PVDC, acrylonitrile and vinyl alcohol copolymers.) Many develoments in the area of compatibility and modified adhesive lamina are being pursued.

NEW MATERIALS

Several new nylon polymers have been developed that exhibit excellent adhesion to polyolefin. These resins can be used in both cast or blown film processes. Allied Chemical's tests indicate that a 400% increase in the force to delaminate, can be realized when used with ethylene vinylacetate copolymers or ionomers. Presumably the modified nylons are copolymers where the structure of the comonomer is similar to that of acrylic acid, vinyl acetate or vinyl alcohol moieties in the polyethylene copolymers.

---

*See Appendix #13.1.1 #26.

MARKETS & APPLICATIONS

In spite of technical complexity and the high initial cost of coextrusion equipment, the coextrusion process consumes an impressive amount of resin. A Shotland Business Study indicates that resin consumption grew over 300% between 1977 and 1980 – 65,000 metric tons to 210,000 metric tons in the film area, and from 41,000 metric tons to 110,000 metric tons of coextruded sheet (Figure 58). Specific producers such as Crown Zellerbach, American Can, Cryovac and St. Regis are involved in very aggressive product development programs. Crown Zellerbach has doubled their product line since 1977 and 85% of Crown Zellerbach film product sales in 1981 were coextruded.*

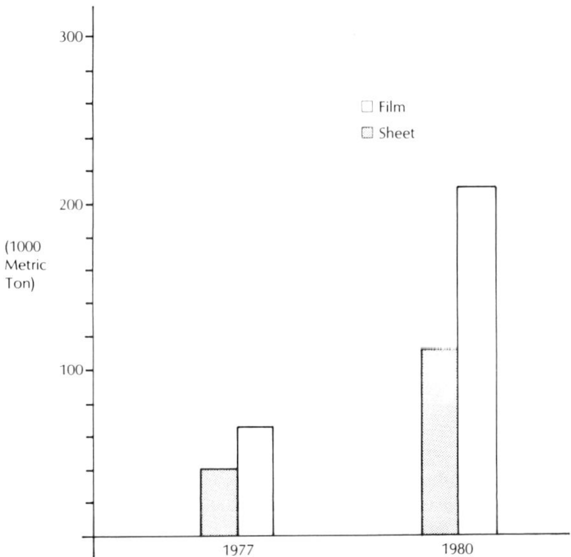

Figure 58. Coextruded film and sheet sales (1000 metric tons) Courtesy of Plastics World.

Coextruded film (or sheet) compositions are widely used in Europe for such items as soft cheese, fruit drinks, yogurt and butter. In the United States these market segments account for only a small fraction of the sales of coextruded products and present a very fertile field.

---

*See Appendix #13.1.1 and #13.1.2-16.

As noted in Modern Plastics February 1981 issue, several market stimuli may increase the use of coextruded films. There are "FDA" mandated dairy products package dating and class A vapor transmission minimums and oxygen transmission regulations for pharmaceutical packaging.

A summary of Coextruded Film Market sales (1979-1981) in the packaging area is contained in Table 48.

Table 49 lists some of the commercial coextruded structures that have replaced the more conventional films.

Table 48. Sales of Coextruded Film Packaging Materials by Market (1000 Metric Ton) (Source: Modern Plastics February 1981)

Market	1979	1981
Trash Bags	58.6	64
Meat Wrap	45.5	59
Snack Packaging	19.5	23
Cheese Packaging	11	13.5
Bakery Overwrap	9	11.9
Cereal Liners	1.9	7
Shipping Sacks	5.9	8
Total	151.4	186.4 (23% growth in 2 years)

Source: Modern Plastics Feb. 1981.

Table 49. Applications of Coextruded Film and Sheet

Application	Coextruded Product
• In-store meat wrap (Replaces PVC and PE)	5 ply shrink film A/EP/EP-PS/A (Layer A = heat seal layer)
• Condiments, puddings, salad dressings	PS/HDPE/PS
• Process meat (Replaces 100% Barex)	MDPE/Barex
• Processed meat (replaces 100% coated N/EVA laminate)	Barrier coated EVAL/N
• Cheese packaging	N/PVAL
• Snack food	HDPE/N/EVA
• Frozen french fry bags	Clear EVA/white LLDPE/clear EVA
• Trash bags (Replaces 2.0 mil LDPE)	LLDPE/LDPE/LLDPE (1.5 mil)
• Kitchen trash bags (Replaces 1.5 LDPE	LLDPE/LDPE/LLDPE (1.3 mil)

Courtesy of Package Engineering, Mar. 1979, p. 39-44.

ECONOMICS

Since the laminar structures are more complex than monolayer films and sheet the ability to calculate costs is equally involved. Some examples of how to calculate the economics or costs of coextruded structures were outlined in a paper presented at a "Coextrusion" Retec in Chicago, June 1981 by Dow Chemical.

The Dow paper notes that some of the added capital and operating costs encountered in coextrusion are those associated with extra extruders, feedblocks, and additional equipment involved in handling more than one polymer. In some cases the ability to use lower cost resins or regrind within the laminar structure more than offsets the higher capital investment and incremental operating costs.

Some conclusions and observations can be made from the Dow data that are not obvious to the casual observer.

The main costs associated with coextrusion vs. monolayer extrusion are:

Extra extruder and feedblock.
Ability to reclaim scrap and reuse, or use lower cost
materials in the core to reduce costs.
Different resin costs and resin handling systems.

In spite of these added expenditures the variations in cost between various coextrusion systems is not excessive. The ability to tailor make a laminar structure in one operation to perform a specific high performance can be visualized as a potential for higher margin products.

Capital cost associated with extrusion and coextrusion lines can vary from $500,000 for a 6″ extruder line to $625,000 for a 6″ 2½″ coextruder line and $770,000 for a 3½″ - 3½″ - 2½″ - 2½″ four extruder coextrusion line.

The output of these lines are 2000 lbs./hour plus or minus 10%. Fixed cost/ lb. of material at capacity varies from 2.1¢ to 3.1¢. Variable cost (utilities etcetera) is 1¢ for both approaches (there are slight variations but they are insignificant).

In four examples the difference between extrusion and coextrusion are just a few cents per pound and (based on 1000 square inches) very minor.

EXAMPLES OF COST ANALYSIS

A. *2 Layer Structure*
Given a monolayer HIPS vs. a 90% HIPS/10% g-p PS coextruded structure, the relative costs associated with 6″ + 2½″ coextrusion line are 6.4¢/lb. for the monolayer and 6.8¢/lb. for the coexruded structure. (Cost of HIPS is 48¢/lb. – cost of HIPS/PS is 47.8¢/lb.) Total cost is 54.4¢/lb. for each film or sheet.

B. *3 Layer Structure*
For monolayer ABS sheet = 87.2¢/lb.
For LDPE/Adhesive/PS = 55.8¢/lb.
(Note: Both systems perform the same function).

C. *5 Layer Structure*
HIPS monolayer is 54.4¢/lb. for coextruded structure consisting of HIPS/adhesive/Saran™/adhesive/HIPS estimated costs 54.9¢/lb.

D. *7 Layer Structure*
For monolayer HIPS = 56¢/lb.
For coextruded structure = 62.6¢/lb.
HIPS/scrap PS/adhesive/Saran™/adhesive/scrap PS/HIPS

It is important to realize that the appropriate layer thickness and extruder sizes are required to produce the lowest cost film. In addition, small savings in raw material costs can easily pay for the extra cost of coextrusion equipment that controls lamina uniformity and thickness. (The use of LLDPE in coextrusion has decreased the cost of coextruded films and allowed significant savings in merchant bags, trash bags and stretch pallet wrap markets.)

For a more detailed treatment of the relative cost of extrusion and coextrusion, the reader is referred to the economic procedures developed by D. E. Clark of Dow Chemical outlined in Modern Plastics October 1981 issue.*

*Appendix #13.1.1 #17.

Chapter 10

Comparison of Commercially Available Shrink Films

The following tables summarize the principal advantages and disadvantages of the main classes of shrink films.

The shortcomings of higher cost, and poor machinability coupled with the overlapping of film properties provide the film manufacturer with a real challenge to improve his films further.

These films will continue to expand their markets and are expected to maintain the 7 to 11% annual growth.

Advantages and Disadvantages of Available Shrink Films:

Table 50 – Vinylidene Chloride/Vinyl Chloride Copolymer
Table 51 – Polyester (PET)
Table 52 – Crosslinked Oriented Polyethylene (LDPE)
Table 53 – Oriented Polyethylene (LDPE)
Table 54 – Polystyrene (OPS)
Table 55 – Oriented Polypropylene (OPP)
Table 56 – Oriented Polyvinylchloride (OPVC)

Table 50. Advantages and Disadvantages of Shrink Films

Film Type	Advantages	Disadvantages
Low Permeability Films		
Vinylidene Chloride/Vinylchloride Copolymer	Low Permeability Suitable for Vacuum Packaging, Odd Shapes Low Shrink and Sealing Temperature	Fair Transparency Machineable on Slow Machines High Cost and Low Area Factor

Table 51. Advantages and Disadvantages of Shrink Films

Film Type	Advantages	Disadvantages
Polyester (PET)	Excellent Mechanical Properties Clarity and Gloss	High Price Non-Sealability except when coated

Table 52. Advantages and Disadvantages of Shrink Films

Film Type	Advantages	Disadvantages
Cross-linked Oriented Polyethylene (LDPE)	Excellent Clarity Low Shrink Temperature Broad Heat-Seal Range Good Tear Resistance and Low-Temperature Properties Good Tack Relatively Low Static Level	Poor Machineability (can be improved by a higher density polyethylene) High Tear Propagation

Table 53. Advantages and Disadvantages of Shrink Films

Film Type	Advantages	Disadvantages
Oriented Polyethylene (LDPE)	Low Price	Poor Machineability Poor Clarity

Table 54. Advantages and Disadvantages of Shrink Films

Film Type	Advantages	Disadvantages
Polystyrene (OPS)	Low Price High O_2 Permeability Sparkle and Clarity	Easily Scratched Brittle

Table 55. Advantages and Disadvantages of Shrink Films

Film Type	Advantages	Disadvantages
Oriented Polypropylene (OPP)	Clarity Low Price and High Yield Good Machineability	Poor Sealability High Shrink Temperature and Tension High Tear Propagation High Static Charge and Dust Attraction

Table 56. Advantages and Disadvantages of Shrink Films

Film Type	Advantages	Disadvantages
Oriented Polyvinyl-chloride (OPVC)	Clarity Reasonable Sealability Medium Shrink Tension Fair Machineability if Moderately Plasticized	Film Relaxation Relatively High Price (because of Low Area Factor)

SHORTCOMINGS OF SHRINK FILMS

Growth of the use of shrink films would be greater were it not for certain shortcomings, the most important of which are cost and machineability.

COST

Except for shrinkable polypropylene, polystyrene, and some grades of polyethylene, shrink films cost considerably more than regular films. (See cost analyses of Wrapping Pallet - Table 57 "Cost Comparision vs. Stretch vs. Competitive Systems for Wrapping Pallet 42″ × 48″.)

MACHINABILITY

Most conventional packaging machines are designed for use with

Table 57. Film Cost Comparision* for Stretch Film vs. Competitive Techniques (Pallet = 48″ × 48″)[1]**

Load Height	Stretch Wrap Spiral	Stretch Wrap Full Web	Shrink Wrap	Steel Strapping*	Poly Strapping*	Cold Glue	Wet (Hot) Glue**	Gaylord Container
30″	$.47	$.39	$ .63	$ .87	$ .82	$.125	$.10	$6.12
40″	.55	.52	.72	.95	.89	.125	.10	6.57
45″	.59	.59	.80	1.02	.92	.125	.10	7.00
50″	.66	.67	.86	1.06	.96	.125	.10	7.50
55″	.68	.72	.94	1.12	1.00	.125	.10	7.65
60″	.72	.78	.98	.1.19	1.05	.125	.10	7.83
65″	.78	.85	1.02	1.22	1.08	.125	.10	8.02
70″	.82	.92	1.10	1.28	1.12	.125	.10	8.40

*Add $.63 to cost of strapping if including edge board.
Assumed 8 units of 12″ edge boards at $.0765 per foot.

**Cost per pallet.

***Add 50% to costs to estimate 1982 costs. (These numbers are 1978-79 actuals. 40% addition may be acceptable for LLDPE cost stretch film —(CJB).

[1]These data originate from raw material and equipment supplies. For similar data see Appendix 13.1.1 #12, 23, 33. and 47 and Appendix 13.1.2 #14, 15, and 16.

Courtesy of "Package Engineering"

cellulose films are not readily converted for the feeding, handling, and heat sealing of synthetic polymer films, especially those that are not dimensionally stable at sealing temperatures.

Chapter 11

Economic Evaluation of Shrink Film Processes and Products

The items involved in total costs, and related to extrusion, are machine and take-off, auxiliary machinery, which consists of dies, supplies, tools, installation, etc., material cost, labor, overhead, and services. A typical example or method of calculating costs are:

AMORTIZED COSTS

Extruder and take-off . . . . (approx.)	$160,000
Auxiliary machinery (dies, supplies, tools, etc.)	40,000
Total fixed investment	$200,000

Assume that payoff is 5 years (with straight-line system of depreciation = $40,000/yr.)

6000 Hours of operation in a year of saleable product . . . . . . $7.00/hr amortization cost of equipment. (When a tenter is required then its cost is included under auxiliary equipment).

MATERIAL COSTS

Assume 400 lb./hr of good product and 45 cents/lb. of compound being fed to the extruder and a loss of 10 lbs./hr. at a salvage worth of 15 cents/lb.

Total material cost becomes = (410 lb. @ 45¢)–(10 lbs. @ 30¢) = $168.0/hr.

LABOR & OVERHEAD COSTS

1 man to run extruder
1/2 man to supply material and work on takeoff
1/2 man to supervise extruder team and maintenance
1 man to run orientation unit

Total = 3 men at $8.00/hr. gross cost = $24.00/hr.

Overhead at 33% salary (standard factor) . . . . $8.00/hr.

SERVICES

Electricity, 50 Kw at 2¢/kw	= $1.00/hr.
Water, air, miscellaneous	= .50/hr
Total	= $1.50/hr.

Manufacturing cost . . . . $700 + $168.00 + $24.00 + $8.00 + $1.50
= $208.50/hr.

at 400 lbs./hr. the cost of operating is 50¢/lb.

This is just an example and each individual case will have its own costs based on equipment, salvage value of scrap, wages, fringes, etc.

Some guidelines that can be used for the various materials and extruder out-puts are listed in the following pages, and Table 58.

One must remember that the operator has to decide the quantity (annual or hourly), that he wants to produce. Ultimately, the best situation is one where the cost per pound of material, or finally the cost per square inch of material, is the lowest.

Table 58. Typical Extrusion Conditions[3]

Product	PS Sheet	PE Tube	PE Flat	PVC Rigid	PVC Plast.
Extruder, HP	100	40	40	25	50
Extruder, Barrel (in/mm)	4½/120	3½/90	3½/90	2½/60	3½/90
Barrel. (Rear Temp.) (°C)	180	150	200	140	140
Barrel. (Frt. Temp.) (°C)	200	160	240	160	170
Die, (Temp.) (°C)	210	165	250	170	175
Die, (Press) (°C)	1050	1500	1000	2250	1500
Melt Temp./ (°C)	210	165	250	175	180
Take-off conditions	Large rolls ready to feed tenter	See Chap. 4.	Chill roll @ 50°C Gap = ½" See Sect. IV.	(1)	(2)

[1]Values for rigid PVC are for a horizontal bubble process.
[2]Values for platicized PVC is for a vertical bubble process.
[3]Temperatures are in degrees centigrade, pressure in psi. Chapter 4 has the details for the processes, and recent machine price quotations are required for accurate figures.

Some typical single-screw extrusion efficiencies have been published, for example:

	lb/hr/hp	**kg/hr/hp**
Rigid PVC	7–10	3 –4.5
Plasticized PVC	10–13	4.5–6
Impact Polystyrene	8–12	3.5–5.5
ABS	5–9	2.3–4
LDPE	7–10	3 –4.5
HDPE	4–8	2 –3.5
Polypropylene	5–10	2.3–4.5

Chapter 12

Future Outlook

GENERAL COMMENTS

Everybody is aware of the political and economic factors encountered in business planning, all of which contribute to the difficulty of producing meaningful forecasts and business plans. However, in spite of these disruptions we can draw conclusions and continue to develop a professional approach to planning based on some definite trends. These disruptions will continue but we believe that our industrial maturity together with our technological base in materials and process technology allows us to predict the most logical paths of growth.

In the 1980's the supply of polymers will be adequate for the packaging industry; the average growth rate of the six leading resins will be an average of 7%/yr., and will remain at this level according to leaders in the industry. Prices will continue to rise. In the specific area of SHRINK AND STRETCH film, the same situation will exist, except that we can expect an annual growth of more than 7%. OPP is expected to exhibit accelerated growth of between 8 and 10%/yr. for the next five years. LDPE and composite laminates using oriented film bases will grow at 8 to 9% minimum. The increased cost of resin will put more pressure on the programs designed to reduce material cost and energy, and increase productivity while at least maintaining margins. THE QUESTION, "SHRINK OR STRETCH?" WILL CONTINUE TO BE ANSWERED ON AN INDIVIDUAL BASIS IN TERMS OF COSTS AND ESTHETICS. This means ¢/in² clarity, cling properties, cost of energy for producing and shrinking, and yield will be important parameters.

Marketing studies have indicated where the growth areas in plastic film markets will occur. For example, indirect packaging (like wraps and bags) will grow 5.2%/yr. through 1995 to 4.5-4.8 billion pounds. This is a little over 1/3 of the total market predicted by 1995. In the December 1981 issue of Plastic World, data were presented from a Predicast Plastic Film Market Study that reinforced some basic concepts that are discussed in detail later.

> Polyethylene film will remain the major workhorse of the plastic packaging films with 2/3 of the total market with major areas of growth in stretch wrap and bags (mostly industrial markets).
> The total film market is expected to grow annually between a pessimistic 4.0% to a repeat of the historical growth of 9-10%. (Realistically, it's somewhere in between.)

There are major global changes coming in polyolefin plastics that will adversely effect the U.S. export market. A drastic drop in U.S. exports of LDPE, HDPE, and PP has been suggested (Figure 58).

The main factors contributing to this decrease are potential new off shore production facilities being built by overseas customers and at the well head.

CANADA will have the most impact with new polyethylene capacity being built in Alberta and British Columbia. 756 million pounds of new LDPE and LLDPE is due on stream in 1985.

100 million pounds of HDPE is due in 1984. (In this instance U.S. price advantage will be eroded, due to low monomer costs in Canada and decontrol of natural gas in the U.S.)

In addition, Canada will bring a new ethylene facility (6.27×10^9 lb/yr capacity) in 1987.

IN THE MIDDLE EAST, Saudia Arabian Basic Industrial Corporation, (SABIC) with Exxon and Shell, will construct a 572 million pound LLDPE, 1.4 million pound ethylene and a 649 million pound styrene complex at Al Jubail.

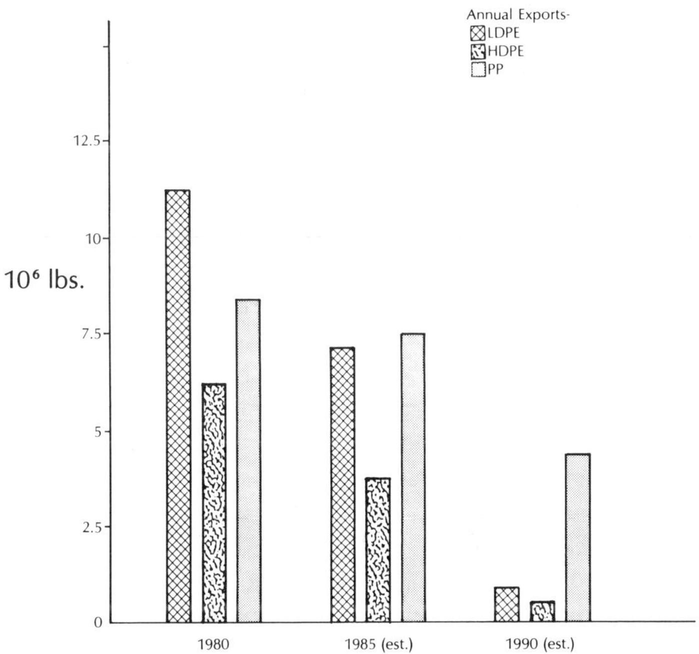

Figure 59. Source: Plastics World Nov. '81, pg 19-21. (Data Source - Chem. Syst. Inc. in a speech to A.C.S./NY. "Mr. B. H. Pickover, Mngr. of Comm'l Dev."). Courtesy of Plastics World.

SABIC and Mobil will build a 1 billion pound ethylene glycol, HDPE, LLDPE complex.

Dow/Mitsubishi/Petrokemya will build a 1.1 billion pound ethylene plant and 396 million pound HDPE and LLDPE facility.

RAW MATERIALS

As noted previously, the price of polyolefins, styrenics, PVC, and polyester will increase but it can be expected that the price of cellophane, coated cellophane, polyester, and aromatics (like styrene polymers) will exhibit price increases greater than the olefins. This is rather evident when one looks at film costs between 1973 and 1978.

New resins will continue to be developed. The one of special note is poly.methylstyrene. Minor modifications will keep developing in specific areas. One such example is a line of heat sealable uniaxially oriented PVC films. Films, produced in Belgium, have a high degree of clarity, gloss, and sparkle. The films have been developed for laminates in the snack food area. Major developments, such as Union Carbide's new line of low pressure, gas phase polymerization LDPE's (recently licensed by Exxon) have touched off a major push to develop alternative routes to low cost processes for LLDPE. The most attractive feature of the new low pressure process is that one can theoretically "swing" the reactors to produce both LLDPE and HDPE. In the shrink and stretch film area there's no expectation of major changes based on this new resin technology.

Publications recently have noted that some executives believe that the price of gas and oil are increasing at a rate that might make it attractive to switch to a coal or vegetable matter base resin technology. On the surface this may look attractive; however, many have not analyzed the alternatives and the technological problems and delays. Coal *can* be a source of raw materials for resin production and the technology is known and has been commercialized in the past. But the alternative of using coal to replace oil as fuel should be thoroughly analyzed. (However, this could conserve capital and aid the balance of trade.)

One suspects that the academicians have not thought out the courses of action required for such a change, i.e. using vegetable or cellulose as the basic polymer building block. Again, on the surface it looks attractive, but cellulose polymers have the wrong structures and would require "major modifications." In addition, the environmental restrictions on new "dissolving pulp" processes could be severe. The major marketer of vegetable matter is the Forest Product Industry. They have spent millions of dollars in research directed at finding a use for lignin. Major projects are also directed at finding use for "pyrolysis oil" (formed from pyrolysis of wood and wood waste). For the next 10 years the vegetable fibers of our forests will be used to generate steam and energy. The Forest Products Industry is dedicated to the objective of becoming energy self-sufficient and if at all possible, they want to be in the co-generation business. There's no reason to believe that new sources (other than gas and oil) will influence the packaging films in the 80's. Unless one finds a low cost cellophane process that is more environmentally innocent.

Raw Material Supply and Cost – Inventories of polyolefins have been adequate and polypropylene is the only resin expected to be in tight supply in the mid-1980's. Prices are rising at a rate that one would expect from rises in oil costs and decontrol of gas. Surging prices of feed stock will continue to be a major consideration for polyolefin products; however, the influx of lower cost LLDPE will probably reduce the rate of growth of polyolefins due to down-gauging and cause serious disruptions in the higher priced LDPE resins.

Producers are "walking a tight rope" and showing concern about:

Voluntary environmental guidelines.

How will the LLDPE effect long term growth and profits of LDPE?

How will one peg his production to a slowing U.S. economy and a decreasing export demand due to the increasing value of the dollar?

CAPACITY AND GROWTH

Several joint European and U.S. surveys have noted that biaxially oriented polypropylene and polyethylene terephthalate dominate the world-wide market for oriented films. In early 1980 world production was shown to be 1.8 billion lbs. In 1979 PET slit film production was estimated to be 900 million to one billion lbs. and oriented polypropylene film approximately 750 to 850 million lbs. In 1978 oriented polypropylene slit film capacity reached 210 million lbs. and OPP is expected to continue its growth by large spurts averaging 8-10%/yr. (one European producer is doubling his capacity in 1980; Northern Petrochem is expected to come on stream with approximately 18 million lbs. of OPP capacity in late 1981 or early 1982). OPP is produced by both tenter and bubble process (sometimes referred to as a double bubble because of the intermediate quench step). The tenter process accounts of 2/3 of the OPP capacity. (As noted in the OPP Section – other companies [CZ, St. Regis, etc.] are also increasing capacity.)

Many published figures concerning oriented film capacity in both Europe and the U.S. are misleading. Several reports estimate the U.S. and European slit film capacity for OPP to be between 195 and 210 million lbs. for the U.S. and 215 to 240 million lbs. for Europe. Japan's strategy is quite evident; their aim is to rely solely on domestic produced OPP and eliminate imported PET to replace cellophane markets.

If one analyzes the cost of producing OPP and relates output to cost of a given process, it can be seen that the cost could decrease from

$2.05/lb./hr to $1.15/lb./hr. conversion cost. This depends almost entirely on energy and size of equipment (leaving room to reduce costs).

For example, a modern 10″ extruder line can produce a 240″ web of OPP at 15 to 18 million lbs./yr. An 8″ line is capable of producing 160″ web at 750 lbs./hr. A 10″ extruder with a chill roll stand MD stretch and tenter can produce 4000 lbs./hr. of OPP (6.4 million lbs./yr. per shift @ 200 working day year). One can easily see from these rough figures that it would certainly pay to increase productivity with the larger units when it takes the same size crew to operate each line.

Polyester, on the other hand, is a more difficult process and demands more complex equipment. Very high quality film is required for the electrical, photographic, and magnetic tape markets. These markets, however, command a higher price than the standard packaging films or laminates.

The present PET and PP markets, (analyzed on the basis of productivity), show that the U.S. averages 50% higher productivity per production line than Europe or Japan, primarily due to larger extruders. *We believe that the large dependence of manufacturing cost on line output will continue to be a deciding factor.* However, attempts to go beyond the 10″ extruder size have exposed some serious problems of heat transfer and cooling the molten material. (The only answer may be two extruders feeding a common stretcher or tenter.) Although exact cost figures are diffiuclt to obtain, it is obvious that the major reason for the substantial decrease of manufacturing cost with size is due essentially to a labor force that is relatively constant with size. Capital investment for the larger units is less/unit output when compared to the smaller lines.

Our research indicates that the lowest manufacturing cost we can calculate accurately for OPP is 1.5 × the RM cost (actually 1.48 times).

To remain the low cost candidate in the shrink film market OPP will require improved product consistency and wider handling characteristics on equipment. To continue its penetration into the cellophane market it will have to maintain its overall cost advantage as a uniform base film, (coated and/or laminated).

On the other hand, PET's cost will continue to be a deterrent to its growing into a large volume food packaging film. Its unique properties and processing characteristics make it an ideal base for engineered packaging materials, (i.e. in laminating, metallizing, coating, etc.).

The film cost of cellophane, polyester and OPP have been increasing since 1973. Cellophane price has increased at an annual rate of 10%/yr.

since 1973. PP on the other hand has averaged 3.5% and PET 4.5%. This is due to the high cost of the energy intensive cellophane process (three to four times the energy required for OPP according to some of the 1979 papers). The need for capital and the environmental restrictions on all new process installations make cellophane expansion and future growth limited.

There is no doubt that OPP will continue to grow due to its price advantage; however, as we noted previously, deeper penetration into select markets will depend on the suppliers ability to control properties and put out a uniform (flat) film. If OPP were stiffer and had better optical properties in laminated form and in coated roll stock, major expansion of the OPP films could be seen in the 80's, far beyond what has been conservatively predicted.

TECHNOLOGY CHANGES

The area of multiple layer co-extrusion holds a lot of potential. Ube Industries in Osaka, Japan has developed a rotational multi-step die that is capable of producing blown film containing the geometric interlocking of two interfacially different polymers. Normally peel strength is low (without an adehsvie layer) but with the Ube process the reports indicate that high peel strengths can be obtained without an adhesive layer. Peel strengths of 1000 gm/in. (TD) to 1500 gm/in. (MD) have been reported with Nylon 6 LDPE at 125 RPM die rotation. Since this new technology produces a less anisotropic composite than the standard blown film dies, one can expect some serious investigations of this technology, in the biaxially oriented film area. Composites like PP/PS, HDPE/Nylon, PP/Nylon could be interesting compositions with two different materials being fed into the die. One serious drawback may be that clear composites with two dissimilar materials is still highly unlikely. However, metallizing could improve the esthetics.

In normal production of OPP the output of the standard extruders appears to be limited by resin capabilities at this time. To significantly increase the output of the present systems (10″ line = 16-18 million lbs./yr.) a polymer melt of special quality or rheology is required. To obtain this rheology one normally is forced to higher molecular weight materials. This puts a severe limitation on the screw design, maximum torque or screw speed. Extruder manufacturers report that they cannot exceed the apparent ceiling of 4000 lbs./hr. with a single screw 10″ extruder.

(Two extruders feeding a common die could be the answer, since most down stream hardware is capable of higher through puts and speeds.) For example, pieces of equipment, such as roll stands, tenters, winders, slitters, etc., can all achieve 1000 ft/min. Cooling efficiency (heat transfer) of chill rolls at these speeds becomes a major problem.

New polyester film lines are being designed for operating speeds of 1200 to 1450 ft./min. with outputs approaching 30 million lbs./yr. If these designs are successful there could be a major push of polyester into the larger volume packaging markets. These developments coupled with technology of high speed coating developed by some of the paper companies, could produce some interesting heat sealable films.

At these high speeds automatic process or computerized process control will become more prevalent. At the present time microprocessors are beginning to be commonplace in many of the plastic converting operations. In OPP, or oriented PET, one can visualize micro-processors being used for controlling temperatures, drive speed, and die adjustments.

The in-line computerized approach allows one to connect the polycondensation (polymerization) directly into film production.

The problems of EPA, OSHA, and the plastic industry will not disappear. (These agencies have not seriously affected the oriented film industry to date.)

Specifically, PET and polyolefin films have not been affected. Oriented PVC has seen some temporary disruptions but this should be short lived, since PVC packaging films have received FDA approval.

Packaging Systems – Film packaging systems will continue to emphasize speed, energy conservation, low cost, and improved performance.

Speed – Speed is being emphasized in number of units packed per time element and in the efficiency of the packaging. Two areas of prominence are the current use of preformed trays with shrink or stretch over-wrap at the fresh meat counter and the use of shrink over-wraps with corrugated trays containing "half cases" ready for "set-up," (bottles, canned goods). Some of the latest cost estimates show that it is now a stand off between a complete corrugated case and a shrink wrapped tray. Some specific items definitely favor the tray approach, especially in the area of packaging from a central distribution station to small stores, or with partial shipments of bulk materials.

A series of high speed tray formers and loaders equipped with facilities for film over-wrap and shrink have been developed. (Mead Packaging's Veri Case™ requires a single operator). Modifications of the L-BAR sealer have also been reported by Packaging Engineering.

Energy – Energy will continue to be a concern. This means that the competitive pressure of stretch film will continue to be felt in the shrink film area and most probably intensify. Shrink tunnels are being redesigned to maximize the energy usage, conserve heat, and produce a package having proepr esthetics with optimum shrink tension. Dupont, Cryovac, Union Carbide and others (Hercules) offer films with improved shrink at less heat.

One such system that is becoming more prevalent is the polyethylene/ethylene-vinylacetate copolymer stretch film that has been modified to provide the proper surface cling without heat sealing. These are being supplied by several companies for bundling. Typical properties are 25% stretch and pre-stretch thicknesses on the order of 1 mil. (Overwrap Equip., Fairfield, CT).

Cost – Cost will continue to be emphasized on a yield and esthetic basis at the retail level. Cryovac D Film (Radiation XLK'D PE) and their coextruded film and bags, Du Pont's Clysar ECL, and EVA copolymers are special materials directed towards giving the optimum physical performance/$.

Orientation can enhance and refine basic product properties to make better use of valuable raw materials. The 1980's will see an every increasing emphasis on expanding this technology to supply the consumer with better food products and items at proper costs.

Chapter 13

Appendices

13.1 BIBLIOGRAPHIES

13.1.1 Recent Packaging Film References
13.1.2 Reference Books, Monographs, and Proceedings from Major Conferences
13.1.3 Historical Patents
13.1.4 European Packaging Applications by Market & Category

13.2 PROPERTIES OF SHRINK AND STRETCH FILMS

13.3 SHRINK AND STRETCH FILM PROPERTIES

13.4 SUPPLIERS OF ORIENTATION EQUIPMENT

13.5 SUPPLIERS OF STRETCH WRAPPING MACHINERY

13.1.1. RECENT PACKAGING FILM ARTICLES

1. *Modern Plastics, January 1982, pg. 16-20.*
2. *"The Top 500 Plastic Processing Plants,"* Plastics World, January 1982, pg. 36-45.
3. *C & EN,* December 21, 1981, pg.37-51 (Markets and Trends)
4. *Modern Plastics,* December 1981, pg. 12-15.
5. J. N. Shirnell, *"Linear Lows," Plastic Technology,* December 1981, pg. 71-75.
6. *"Rating Stretch Film Performance in Plant,"* Packaging Engineering, December 1981, pg. 62-63.
7. *"Supply and General Markets," Plastics World* (April 1981 pg. 78; November 1981 pg. 11-16).
8. *"How to describe the outlook for PE 'Creative Chaos' is no exaggeration,"* Modern Plastics, November 1981, pg. 63-65.
9. *Barge Mounted PE Plant Starts Up,* Modern Plastics, November 1981, pg. 16.
10. *Modern Plastics,* November 1981, pg. 18-20 *"Alaska - Next Big Producer of Monomer and Resin?*
11. *"Surface Treatment Improves Polyethylene Barrier Properties,"* Package Engineering, November 1981, pg. 64-66.
12. *"Stretch Wrapping Pallet Loads"* – Anon. Packaging Engineering, November 1981 pg. 48-50.
13. D. E. McGillan, *"Quality Control of Flexible Packaging,"* Packaging Engineering November 1981, pg. 56-60.
14. *"Palletizing Bundling Ideas Respond to Productivity Needs,"* Package Engineering, August 1981, pg. 27-28.
15. *C & EN,* November 30, 1981, pg. 13-14 (Key Chemicals).
16. *Concentrates for LLDPE and other Thermoplastics,* Plastic World, October 1981, pg. 104.
17. C. R. Finch, *"Coextrusion Economics: How to Calculate Them,"* Modern Plastics, October 1981, pg. 68-71.
18. *"Concentrates for LLDPE and other Thermoplastics,"* Plastics World, October 1981, pg. 104.
19. *Modern Plastics,* October 1981, pg. 20.
20. Mario Concha *"LLDPE – The New Billion Pound Resins,"* 7th Annual "Planning with Plastics" Conference N.Y.C., September 1981.
21. *Plastic Producers Search for Paths to LLDPE,* Chemical Engineering, August 24, 1981, pg. 47, 48.

22. *"Anti Slip Wrap Adds Stability to Loads"* – Anon. Package Engineering, August 1981, pg. 30.
23. *"Stretch, Shrink Bundle Options"* – Anon. Packaging Engineering, August 1981 pg. 31.
24. *Package Engineering,* August 1981, pg. 41-44 *"LLDPE Blends New Worth into Film."*
25. R. Bloor, *"New Way to Crosslink P.E.,"* Plastics Technology, February 1981, pg. 83-86.
26. J. Sneller, *"High Performance Coextrusion,"* Modern Plastics, February 1981, pg. 36-39.
27. M. Harting, *"Running the Linear Lows," Plastic Technology,* February 1981.
28. W. A. Fraser, L. S. Sarola and M. Concha, *"Film Extrusion of LP-LDPE,"* Polyolefin Div'n. UC&C. 1980 Proceedings SPE Conf.
29. Adhesives Age, November 1980 (Predicast Data).
30. *"High Density Polyethylene,"* Modern Plastics, Morsine and Sneller, August 1979.
31. *"Oriented Polypropylene...Metallization,"* P. Prince, J. Scharr, Package Engineering, August 1979, pg. 605-606.
32. *"EVA 3135X Dupont,"* Package Engineering, July 1979, pg. 37.
33. *"Stretch Film"* – Anon. Package Engineering, July 1979, pg. 38-41.
34. Ericson, J. E., *"Polyester Films for Packaging Applications,"* Preprints 37th ANTEC: S.P.E. May 7-10, 1979 pg. 718-720.
35. *Package Engineering,* April 1979 pg. 49-51 *Packaged Meats,* H. Daun & Seymour G. Gilbert.
36. Andres, C. *"Vacuum Packaging Equipment Key to Central Meat processing,"* Food Processing 36:2, pg. 61, 1979.
37. *"Shrink or Stretch Equipment,"* Martin Siegel, Modern Packaging Encyclopedia, December 1978 (pg. 145-148).
38. *"Shrink Packaging...Performance"* – Anon Package Enginering, September 1978, pg. 38.
39. *"The Challenge of Meat, Modern Packaging,* August 1977.
40. *"Helping Bakery Profits Rise,"* Modern Packaging, March 1977.
41. *"Effect of Decreasing Temperature & Polyethylene Wraps on Ripening and Respiration of Avocado Fruit"* Jour, Israel Agr. Res. 18: (2) 77 by Aharoni, Y. M.
42. Hartung, A. *"Production of Low Density Polyethylene Shrink Film."* Kunststoffe 67, 3, 126-9 1977.
43. *"What Shrink Filmn is...,"* M. Klauber, Package Development, May/-June 1975.

44. *"Stretch or Shrink,"* R. Martino, Modern Platics, February 1975, pg. 49-52.
45. Gensinger, D. *"Shrink Films for Dairy Industry"* Milchwirtsch Ber. Bundesanst Wolfpassing Rotholz 43, 145-8 1975.
46. *"Study of Resin Types,"* D. Rex Young, Package Development, November-December 1974, pg. 24-27.
47. *"Wrapping Things in Plastic,"* J. Hager, D. Rex Young, *Chemtech,* November-December 1974, pg. 24–27.
48. Klauber, M. *"Simplified Test for High Temperature Strength of PE Shrink Film"* Mod. Plast. 50, 2, 74-6 1973.
49. Miller, R. W., *"Polypropylene Film in Packaging,"* S.P.E. ANTEC 1973.
50. Pirot, E. *"Extrusion of Oriented Tube Films (Shrink Films)* Ver. Deut. Ing. Z, 11, 2, 155-9 1972.
51. Pirot, E. *"Production of PVC Shrink Films"* Kunststoffe 61, 4, 222-225 1971.

13.1.2 REFERENCE BOOKS, MONOGRAPHS AND PROCEEDINGS OF MAJOR CONFERENCES

1. J. H. Briston *"Plastic Films"* Halsted Press, John Wiley & Sons, NY.
2. *Science and Technology of Polymer Films* Ed. by Orville J. Sweeting, Interscience Publishers div. of John Wiley & Sons.
3. *Plastics Films & Packaging* by C. R. Oswin, Halsted Press, John Wiley & Sons.
4. Coveney, R. D. & Jones, A. Proc. Pack. Pres. Food Congress (Harrogate, *1969*) Leatherhead PIRA.
5. J. H. Briston and L. I. Katan, *Plastics in Contact with Foods,* London Food Trade Press, 1974.
6. Sarvetnik, H. A. *Polyvinyl Chloride,* NY Reinhold.
7. Rubin, I. D. "Poly(l-butene) its Preparation & Properties, Polymer Monographs, Gordon & Breach Science Publishers, NY 1968.
8. R. H. Boundy and R. F. Boyer, "Styrene, Its Polymers, Copolymers and Derivatives," Reinhold Publishing Corp., New York.
9. A. V. Tobolsky, *Properties and Structure of Polymers,* Wiley, NY 1960.
10. American Society of Horticulture Science. Annual Proceedings 1974→1981.
11. *Proceedings of M R I Symposium No. 3 "Meat Freezing: Why & How* -1974.
12. Michigan State Univ. Jan. 27 & 29,1969; Proc. of the Nat'l Cont. Atm. Res. Conf.
13. TAPPI, "Paper Synthetics" proceedings - 1974 to 1981.
14. *Modern Packaging* Encyclopedia December 1982.
15. *Packaging Encyclopedia* – 1982 Edition.
16. *Plastics World* – 1982 Plastics Directory (Vol. 40/No. 2).
17. Modern Plackaging Films, Ed. by S. H. Pinner, Publ. by Butterworths (London), 1967.

13.1.3 HISTORICAL PATENTS

1. Kaiser, U. H. "Production of Polyethylene Shrink Films" Kunststoffe 63, 8, 505-6 1973.
2. Japanese Patent 7390367 "Shrink Resistant Stretched PVC Films" Kawaguchi, O. assigned to Mitsubishi Plastic Industries Ltd. Public date Nov. 26, 1973.
3. South African Patent 7101642 "Laminated Plastic Film for Heat-Shrink Packaging" Schirmer, H. G. (Public date 29 Nov. 71).
4. C. A. Lindstrom et al (to W. R. Grace) Can. Pat. 639,424 (1962).
5. W. G. Baird et al (to W. R. Grace) U.S. Pat. 3,022,543 (1962).
6. E. R. Winter (to DuPont) U.S. Pat. 2,995,779 (1961).
7. R. K. Burkhard et al (to Reynolds Metals Co.) U.S. Pat. 3,039,147 (1961).
8. W. G. Baird et al (to W. R. Grace) Brit. Pat. 849,070 (1960).

13.1.4 EUROPEAN PACKAGING APPLICATIONS BY MARKET AND CATEGORY

Milk	85% Glass (81% returnable)	8% Metal	2% LDPE	5% waxed or PE coated cartons
Carbonated Beverages	85% Glass (70% returnable)	12% Metal	3% laminates and pouches (PET and LDPE)	
Processed Vegetables	5% Glass	94% Metal	1% (foil/PET/LDPE)	
Processed Meat	3-5% Glass	92-95% Metal	3% (LDPE/Nylon)	
Fresh Meat	–	–	Stretch PVC over pulp or PS Trays	
Poultry	–	–	LDPE;EVAC	
Fresh Vegetables	–	–	LDPE, PS, PVC COPOLYMERS	
Fruit	–	–	LDPE, PS, PVC	
Snack Foods	–	–	OPP (Coated, coextruded or laminated)	
Biscuits	–	–	OPP (Coated) and Wax laminated paper	
Bread	–	–	LDPE and Waxed paper	
Cookies	–	–	PVC and OP thermoformed trays and overwrapped with coated OPP	
Candy	–	–	Coated cast PP, LDPE Coated oriented PP Vacuum formed OPS trays PVC thermoformed trays	

SECTION 13.2 SHRINK AND STRETCH FILM PROPERTIES

Types	Gauge (um/mil)	Tensile (psi ×10³)	WVTR (gm/mil)	Oxygen (c.c./mil)	Shrink (max.%)	Tension (psi)	Temp. (F)	Sealing (temp. F)
Stretch Pliofilm	10/04	6-19	11-21	90	40-50	150-350	150-230	180-250
Polybutene-1	12½-5/.5-2.0	17-21	8-10	1500-3000	40-80	100-350	190-350	300-400
Polyester	12½/0.5	24-36	15	80-120	45-55	700-1600	160-250	——
Stretch PE	25-50/1-2	1.6-3.1	13-18	5600	20-75	50-100	190-300	250-
Stretch PE/EVA	25-75/1-3	2-3.5	15	11,000	20-70	40-90	150-250	200-350
Shrink PE	25-50/1-2	9	10-15	5,000	80	200-500	190-280	230-400
Shrink Rad xlk PE	-/.6-1.5	8-19	5-10	5,000	50-80	250-500	160-290	300-500
OPP	12-37/.5-1.5	15-27	5	2,000	50-80	300-600	200-350	350-400
OPA	25/1.0	9-12	60+	3,500	40-70	100-600	210-270	250-300
Stretch PVC	12-37/.5-1.5	10-20	50+	300-8,000	30-70	150-300	150-300	275-370
Shrink & Stretch Saran	/.4-1.0	7-20	3-9	15-40	15-60	50-200	140-290	250-300

NOTE:
WVTR = gm/m²/24 hrs./mil @ 100 F(38C)- 90% RH
OXYGEN = c.c./m²/24 hrs./mil. @ R.T. - i atm. (15 psi)
MAXIMUM SHRINK determined in boiling water or Silicone Oil at 100C for 5 seconds immersion (W. R. Grace Data)
SHRINK TENSION DETERMINED AT SAME CONDITIONS AS MAXIMUM SHRINK (using strain gauges in frame cont. 1″ × 3″ film strips).

13.3. SHRINK AND STRETCH FILM SUPPLIERS (US)

Source	PE & EVA Stretch	PE Shrink	OPP	OPS	PVC	Poly-ester	Saran
A & E Plastics				X			
Alliance Paper (Brentwood, NY)	X	X	X	X	X	X	X
Allied Chemical					X		
Amer. Can Co.	X						
Amer. Hoechst Co.					X	X	
Arco Polymers	X		X				
Ashland Oil	X						
Baker & Malcolm	X				X		
Bemis Co.	X		X				
Blako Ind.	X						
Boyertown Pack.	X						
Cadillac Plastics	X	X	X	X	X	X	
Cellu-Craft Inc.	X						
Central Bag (CA)X			X				
Chase Bag	X						
Chemplex	X						
Cleve. Plas. Films	X						
Clopay Corp.					X		
Continental Can	X						
Crown Zellerbach	X		X				X
Cryovac Div. (WR Grace)		X	X	X			
Dayco Corp.					X		X
Diamond Shamrock	X				X		
Dow Chemical				X			X
Dupont De Nemours	X	X				X	
Dura-Lee Corp.	X		X		X		
Edison Plastics	X		X				

(Continued)

Source	PE & EVA Stretch	PE Shrink	OPP	OPS	PVC	Poly-ester	Saran
Errich Packaging	X		X				X
Ethyl Corp.	X				X		
Exxon Chem.	X		X				
Favorite Plastics	X						
F M C	X						X
Flex-O-Film				X			
Goodyear Tire & Rub.	X		X	X	X	X	
Hercules Inc.		X					
ICI	X	X				X	
Kimball Container	X				X		
Kleerpak	X		X		X		
Larson Packaging	X						X
Liqui-Box Corp.	X						
Metal Box	X						
Mir-Pak Inc.	X		X				
Mitsubishi				X	X	X	
Mobil Chem.	X			X			
Monsanto				X			
Northern Petro	X						
PCM Inc.	X		X				
Package Printing	X		X	X			
PPD Co.	X						
Research Pack. Ser.	X		X		X		
Resinite Div. (Borden)	X						
Reynolds Metals					X		
St. Regis Paper	X		X		X	X	X
Sealcraft Prod.	X		X				
Shield Pack Inc.	X		X				
Tee-Pack							X
3M Co.						X	
Tri-State Pack.	X		X				X

(Continued)

Source	**PE & EVA Stretch**	**PE Shrink**	**OPP**	**OPS**	**PVC**	**Poly-ester**	**Saran**
UCB/SIDAC	X		X				X
Union Camp Corp.	X						
Union Carbide	X	X		X	X		
US Packaging Corp.	X		X				X
US Steel	X				X		
Weldotron Corp.	X	X	X				X
Wilder Industries	X		X	X		X	X

PLEASE NOTE: A COMPREHENSIVE AND UP-DATED LIST OF THOSE WHO SUPPLY SHRINK FILMS CAN BE FOUND IN THE "MODERN PLASTICS ENCYCLOPEDIA - 1982 191-82 BUYERS GUIDE; 1982 PLASTIC DIRECTORY (PLASTICS WORLD VOL 40/NO 2); PACKAGING ENGINEERING - 1982 ENCYCLOPEDIA

13.4 SUPPLIERS OF ORIENTATION EQUIPMENT

Aidlin Automation Inc., Brooklyn, NY
Black Clawson Co., Fulton, NY
Lyons Machine Works, Paterson, NJ
Marshall & Williams Co., Providence, RI
Proctor & Schwartz, Philadelphia, PA
Reifenhauser, Saddle Brook, NJ
Ringler-Dorin Inc., East Farmingdale, NY
Worldwide Converting Machines, Allendale, NJ

This is a partial list. For a more complete and comprehensive list the reader is referred to Modern Plastics Encyclopedia, the 1981-1982 Annual Buyers Guide or 1982 Plastics Directory (Plastics World - Vol. 40/No. 4).

(Note: manufacturers of extrusion-blown film lines normally produce orientation equipment on an "as Specified" basis for their customers. These production lines are for the "bubble process" for orientation, rather than the tenters emphasized above.)

13.5 SUPPLIERS OF STRETCH WRAPPING MACHINERY

Arenco Machine Co.
Consolidated Thermoplastics
Liqui-Box Corp.
Lantech Inc.
Union Camp Packaging Systems
Allied Automation
Weldotron Corpo.
Stone Container Corp.
Favorite Plastics
Interstate Inc.
Lithibar Co.
International Packaging Machinery
Packaging Sales & Dev.
Trident Manufacturing Co.
Overwrap Equipment Corp.
Infra-Pak

This is a partial list. For a more complete and comprehensive list the reader is referred to Modern Plastics Encyclopedia, the 1981-1982 Annual Buyers Guide, or 1982 Plastics Directory (Plastics World Vol. 40/No. 4).